INTRODUCTION TO OPTICAL WAVEGUIDE ANALYSIS

INTRODUCTION TO OPTICAL WAVEGUIDE ANALYSIS
Solving Maxwell's Equations and the Schrödinger Equation

KENJI KAWANO and TSUTOMU KITOH

A Wiley-Interscience Publication
JOHN WILEY & SONS, INC.
New York / Chichester / Weinheim / Brisbane / Singapore / Toronto

This book is printed on acid-free paper. ∞

Copyright © 2001 by John Wiley & Sons. All rights reserved.

Published simultaneously in Canada.

No part of this publication may be reproduced, stored in a retrieval system or transmitted in any form or by any means, electronic, mechanical, photocopying, recording, scanning or otherwise, except as permitted under Section 107 or 108 of the 1976 United States Copyright Act, without either the prior written permission of the Publisher, or authorization through payment of the appropriate per-copy fee to the Copyright Clearance Center, 222 Rosewood Drive, Danvers, MA 01923, (978) 750-8400, fax (978) 750-4744. Requests to the Publisher for permission should be addressed to the Permissions Department, John Wiley & Sons, Inc., 605 Third Avenue, New York, NY 10158-0012, (212) 850-6011, fax (212) 850-6008, E-Mail: PERMREQ@WILEY.COM.

For ordering and customer service, call 1-800-CALL-WILEY.

Library of Congress Cataloging-in-Publication Data:

Kawano, Kenji.
   [Hikari doharo kaiseki no kiso. English]
   Introduction to optical waveguide analysis: solving Maxwell's equations and the Schrödinger equation/by Kenji Kawano, Tsutomu Kitoh.
     p. cm.
   ISBN 0-471-40634-1 (cloth)
   1. Optical waveguides—Mathematical models. 2. Maxwell equations. 3. Schrödinger equation. I. Kitoh, Tsutomu. II. Title.

TAI1750.K3913 2001
621.381′331—dc21   2001017920

Printed in the United States of America

10 9 8 7 6 5 4 3 2 1

To our wives,
Mariko and Kumiko

CONTENTS

Preface / xi

1 Fundamental Equations   1

1.1 Maxwell's Equations / 1
1.2 Wave Equations / 3
1.3 Poynting Vectors / 7
1.4 Boundary Conditions for Electromagnetic Fields / 9
Problems / 10
Reference / 12

2 Analytical Methods   13

2.1 Method for a Three-Layer Slab Optical Waveguide / 13
2.2 Effective Index Method / 20
2.3 Marcatili's Method / 23
2.4 Method for an Optical Fiber / 36
Problems / 55
References / 57

3 Finite-Element Methods 59

3.1 Variational Method / 59
3.2 Galerkin Method / 68
3.3 Area Coordinates and Triangular Elements / 72
3.4 Derivation of Eigenvalue Matrix Equations / 84
3.5 Matrix Elements / 89
3.6 Programming / 105
3.7 Boundary Conditions / 110
Problems / 113
References / 115

4 Finite-Difference Methods 117

4.1 Finite-Difference Approximations / 118
4.2 Wave Equations / 120
4.3 Finite-Difference Expressions of Wave Equations / 127
4.4 Programming / 150
4.5 Boundary Conditions / 153
4.6 Numerical Example / 160
Problems / 161
References / 164

5 Beam Propagation Methods 165

5.1 Fast Fourier Transform Beam Propagation Method / 165
5.2 Finite-Difference Beam Propagation Method / 180
5.3 Wide-Angle Analysis Using Padé Approximant Operators / 204
5.4 Three-Dimensional Semivectorial Analysis / 216
5.5 Three-Dimensional Fully Vectorial Analysis / 222
Problems / 227
References / 230

6 Finite-Difference Time-Domain Method 233

6.1 Discretization of Electromagnetic Fields / 233
6.2 Stability Condition / 239
6.3 Absorbing Boundary Conditions / 241

Problems / 245
References / 249

7 Schrödinger Equation 251

7.1 Time-Dependent State / 251
7.2 Finite-Difference Analysis of Time-Independent State / 253
7.3 Finite-Element Analysis of Time-Independent State / 254
References / 263

Appendix A Vectorial Formulas 265

Appendix B Integration Formula for Area Coordinates 267

Index 273

PREFACE

This book was originally published in Japanese in October 1998 with the intention of providing a straightforward presentation of the sophisticated techniques used in optical waveguide analyses. Apparently, we were successful because the Japanese version has been well accepted by students in undergraduate, postgraduate, and Ph.D. courses as well as by researchers at universities and colleges and by researchers and engineers in the private sector of the optoelectronics field. Since we did not want to change the fundamental presentation of the original, this English version is, except for the newly added optical fiber analyses and problems, essentially a direct translation of the Japanese version.

Optical waveguide devices already play important roles in telecommunications systems, and their importance will certainly grow in the future. People considering which computer programs to use when designing optical waveguide devices have two choices: develop their own or use those available on the market. A thorough understanding of optical waveguide analysis is, of course, indispensable if we are to develop our own programs. And computer-aided design (CAD) software for optical waveguides is available on the market. The CAD software can be used more effectively by designers who understand the features of each analysis method. Furthermore, an understanding of the wave equations and how they are solved helps us understand the optical waveguides themselves.

Since each analysis method has its own features, different methods are required for different targets. Thus, several kinds of analysis methods have to be mastered. Writing formal programs based on equations is risky unless one knows the approximations used in deriving those equations, the errors due to those approximations, and the stability of the solutions.

Mastering several kinds of analysis techniques in a short time is difficult not only for beginners but also for busy researchers and engineers. Indeed, it was when we found ourselves devoting substantial effort to mastering various analysis techniques while at the same time designing, fabricating, and measuring optical waveguide devices that we saw the need for an easy-to-understand presentation of analysis techniques.

This book is intended to guide the reader to a comprehensive understanding of optical waveguide analyses through self-study. It is important to note that the intermediate processes in the mathematical manipulations have not been omitted. The manipulations presented here are very detailed so that they can be easily understood by readers who are not familiar with them. Furthermore, the errors and stabilities of the solutions are discussed as clearly and concisely as possible. Someone using this book as a reference should be able to understand the papers in the field, develop programs, and even improve the conventional optical waveguide theories.

Which optical waveguide analyses should be mastered is also an important consideration. Methods touted as superior have sometimes proven to be inadequate with regard to their accuracy, the stability of their solutions, and central processing unit (CPU) time they require. The methods discussed in this book are ones widely accepted around the world. Using them, we have developed programs we use on a daily basis in our laboratories and confirmed their accuracy, stability, and effectiveness in terms of CPU time.

This book treats both analytical methods and numerical methods. Chapter 1 summarizes Maxwell's equations, vectorial wave equations, and the boundary conditions for electromagnetic fields. Chapter 2 discusses the analysis of a three-layer slab optical waveguide, the effective index method, Marcatili's method, and the analysis of an optical fiber. Chapter 3 explains the widely utilized scalar finite-element method. It first discusses its basic theory and then derives the matrix elements in the eigenvalue equation and explains how their calculation can be programmed. Chapter 4 discusses the semivectorial finite-difference method. It derives the fully vectorial and semivectorial wave equations, discusses their relations, and then derives explicit expressions for the quasi-TE and quasi-TM modes. It shows formulations of E_x and H_y expressions for the quasi-TE (transverse electric) mode and E_y and H_x expressions for the quasi-TM (transverse magnetic) mode. The nonequidistant discretization scheme used in this chapter is more versatile than the equidistant discretization reported by Stern. The discretization errors due to these formulations are also discussed. Chapter 5 discusses beam propagation methods for the design of two- and three-dimensional (2D, 3D) optical waveguides. Discussed here are the fast Fourier transform beam propagation method (FFT-BPM), the finite-difference beam propagation method (FD-BPM), the transparent boundary conditions, the wide-angle FD-BPM using the Padé approximant operators, the 3D semivectorial analysis based on the alternate-direction implicit method, and the fully vectorial analysis. The concepts of these methods are discussed in detail and their equations are derived. Also discussed are the error factors of the FFT-BPM, the physical meaning of the Fresnel equation, the problems with the wide-angle FFT-BPM, and the stability of the FD-BPM. Chapter 6 discusses the finite-difference time-domain method (FD-TDM). The FD-TDM is a little difficult to apply to 3D optical waveguides from the viewpoint of computer memory and CPU time, but it is an important analysis method and is applicable to 2D structures. Covered in this chapter are the Yee lattice, explicit 3D difference formulation, and absorbing boundary conditions. Quantum wells, which are indispensable in semiconductor optoelectronic devices, cannot be designed without solving the Schrödinger equation. Chapter 7 discusses how to solve the Schrödinger equation with the effective mass approximation. Since the structure of the Schrödinger equation is the same as that of the optical wave equation, the techniques to solve the optical wave equation can be used to solve the Schrödinger equation.

Space is saved by including only a few examples in this book. The quasi-TEM and hybrid-mode analyses for the electrodes of microwave integrated circuits and optical devices have also been omitted because of space limitations. Finally, we should mention that readers are able to get information on the vendors that provide CAD software for the numerical methods discussed in this book from the Internet.

We hope this book will help people who want to master optical waveguide analyses and will facilitate optoelectronics research and development.

KENJI KAWANO and TSUTOMU KITOH

Kanagawa, Japan
March 2001

INTRODUCTION TO OPTICAL WAVEGUIDE ANALYSIS

CHAPTER 1

FUNDAMENTAL EQUATIONS

This chapter summarizes Maxwell's equations, vectorial wave equations, and the boundary conditions for electromagnetic fields.

1.1 MAXWELL'S EQUATIONS

The electric field **E** (in volts per meter), the magnetic field **H** (amperes per meter), the electric flux density **D** (coulombs for square meters), and the magnetic flux density **B** (amperes per square meter) are related to each other through the equations

$$\mathbf{D} = \varepsilon \mathbf{E}, \tag{1.1}$$

$$\mathbf{B} = \mu \mathbf{H}, \tag{1.2}$$

where the permittivity ε and permeability μ are defined as

$$\varepsilon = \varepsilon_0 \varepsilon_r, \tag{1.3}$$

$$\mu = \mu_0 \mu_r. \tag{1.4}$$

Here, ε_0 and μ_0 are the permittivity and permeability of a vacuum, and ε_r and μ_r are the relative permittivity and permeability of the material. Since the relative permeability μ_r is 1 for materials other than magnetic materials, it is assumed throughout this book to be 1. Denoting the velocity of light in a vacuum as c_0, we obtain

$$\varepsilon_0 = \frac{1}{c_0^2 \mu_0} \approx 8.854188 \times 10^{-12} \quad \text{F/m} \tag{1.5}$$

$$\mu_0 = 4\pi \times 10^{-7} \quad \text{H/m}. \tag{1.6}$$

The current density **J** (in amperes per square meter) in a conductive material is given by

$$\mathbf{J} = \sigma \mathbf{E}. \tag{1.7}$$

The electromagnetic fields satisfy the following well-known Maxwell equations [1]:

$$\nabla \times \mathbf{E} = -\frac{\partial \mathbf{B}}{\partial t}, \tag{1.8}$$

$$\nabla \times \mathbf{H} = \frac{\partial \mathbf{D}}{\partial t} + \mathbf{J}. \tag{1.9}$$

Since the equation $\nabla \cdot (\nabla \times \mathbf{A}) = 0$ holds for an arbitrary vector **A**, from Eqs. (1.8) and (1.9), we can easily derive

$$\nabla \cdot \mathbf{B} = 0, \tag{1.10}$$

$$\nabla \cdot \mathbf{D} = \rho. \tag{1.11}$$

The current density **J** is related to the charge density ρ (in coulombs per square meter) as follows:

$$\nabla \cdot \mathbf{J} = -\frac{\partial \rho}{\partial t}. \tag{1.12}$$

Equations (1.10) and (1.11) can be derived from Eqs. (1.8), (1.9), and (1.12).

1.2 WAVE EQUATIONS

Let us assume that an electromagnetic field oscillates at a single angular frequency ω (in radians per meter). Vector **A**, which designates an electromagnetic field, is expressed as

$$\mathbf{A}(\mathbf{r}, t) = \text{Re}\{\bar{\mathbf{A}}(\mathbf{r}) \exp(j\omega t)\}. \tag{1.13}$$

Using this form of representation, we can write the following phasor expressions for the electric field **E**, the magnetic field **H**, the electric flux density **D**, and the magnetic flux density **B**:

$$\mathbf{E}(\mathbf{r}, t) = \text{Re}\{\bar{\mathbf{E}}(\mathbf{r}) \exp(j\omega t)\}, \tag{1.14}$$

$$\mathbf{H}(\mathbf{r}, t) = \text{Re}\{\bar{\mathbf{H}}(\mathbf{r}) \exp(j\omega t)\}, \tag{1.15}$$

$$\mathbf{D}(\mathbf{r}, t) = \text{Re}\{\bar{\mathbf{D}}(\mathbf{r}) \exp(j\omega t)\}, \tag{1.16}$$

$$\mathbf{B}(\mathbf{r}, t) = \text{Re}\{\bar{\mathbf{B}}(\mathbf{r}) \exp(j\omega t)\}. \tag{1.17}$$

In what follows, for simplicity we denote $\bar{\mathbf{E}}$, $\bar{\mathbf{H}}$, $\bar{\mathbf{D}}$, and $\bar{\mathbf{B}}$ in the phasor representation as **E**, **H**, **D**, and **B**. Using these expressions, we can write Eqs. (1.8) to (1.11) as

$$\nabla \times \mathbf{E} = -j\omega \mathbf{B} = -j\omega \mu_0 \mathbf{H}, \tag{1.18}$$

$$\nabla \times \mathbf{H} = j\omega \mathbf{D} = j\omega \varepsilon \mathbf{E}, \tag{1.19}$$

$$\nabla \cdot \mathbf{H} = 0, \tag{1.20}$$

$$\nabla \cdot (\varepsilon_r \mathbf{E}) = 0, \tag{1.21}$$

where it is assumed that $\mu_r = 1$ and $\rho = 0$.

1.2.1 Wave Equation for Electric Field E

Applying a vectorial rotation operator $\nabla \times$ to Eq. (1.18), we get

$$\nabla \times (\nabla \times \mathbf{E}) = -j\omega \mu_0 \nabla \times \mathbf{H}. \tag{1.22}$$

Using the vectorial formula

$$\nabla \times (\nabla \times \mathbf{A}) = \nabla(\nabla \cdot \mathbf{A}) - \nabla^2 \mathbf{A}, \qquad (1.23)$$

we can rewrite the left-hand side of Eq. (1.22) as

$$\nabla(\nabla \cdot \mathbf{E}) - \nabla^2 \mathbf{E}. \qquad (1.24)$$

The symbol ∇^2 is a Laplacian defined as

$$\nabla^2 = \frac{\partial^2}{\partial x^2} + \frac{\partial^2}{\partial y^2} + \frac{\partial^2}{\partial z^2}. \qquad (1.25)$$

Since Eq. (1.21) can be rewritten as

$$\nabla \cdot (\varepsilon_r \mathbf{E}) = \nabla \varepsilon_r \cdot \mathbf{E} + \varepsilon_r \nabla \cdot \mathbf{E} = 0,$$

we obtain

$$\nabla \cdot \mathbf{E} = -\frac{\nabla \varepsilon_r}{\varepsilon_r} \cdot \mathbf{E}. \qquad (1.26)$$

Thus, the left-hand side of Eq. (1.22) becomes

$$-\nabla \left(\frac{\nabla \varepsilon_r}{\varepsilon_r} \cdot \mathbf{E} \right) - \nabla^2 \mathbf{E}. \qquad (1.27)$$

On the other hand, using Eq. (1.19), we get for the right-hand side of Eq. (1.22)

$$k_0^2 \varepsilon_r \mathbf{E}, \qquad (1.28)$$

where k_0 is the wave number in a vacuum and is expressed as

$$k_0 = \omega \sqrt{\varepsilon_0 \mu_0} = \frac{\omega}{c_0}. \qquad (1.29)$$

Thus, for a medium with the relative permittivity ε_r, the vectorial wave equation for the electric field **E** is

$$\nabla^2 \mathbf{E} + \nabla\left(\frac{\nabla \varepsilon_r}{\varepsilon_r} \cdot \mathbf{E}\right) + k_0^2 \varepsilon_r \mathbf{E} = \mathbf{0}. \tag{1.30}$$

And using the wave number k in that medium, given by

$$k = k_0 n = k_0 \sqrt{\varepsilon_r} = \omega\sqrt{\varepsilon_0 \varepsilon_r \mu_0} = \omega\sqrt{\varepsilon \mu_0}, \tag{1.31}$$

we can rewrite Eq. (1.30) as

$$\nabla^2 \mathbf{E} + \nabla\left(\frac{\nabla \varepsilon_r}{\varepsilon_r} \cdot \mathbf{E}\right) + k^2 \mathbf{E} = \mathbf{0}. \tag{1.32}$$

When the relative permittivity ε_r is constant in the medium, this vectorial wave equation can be reduced to the Helmholtz equation

$$\nabla^2 \mathbf{E} + k^2 \mathbf{E} = \mathbf{0}. \tag{1.33}$$

1.2.2 Wave Equation for Magnetic Field H

Applying the vectorial rotation operator $\nabla \times$ to Eq. (1.19), we get

$$\nabla \times (\nabla \times \mathbf{H}) = j\omega\varepsilon_0 \nabla \times (\varepsilon_r \mathbf{E}).$$

Thus,

$$\begin{aligned}\nabla(\nabla \cdot \mathbf{H}) - \nabla^2 \mathbf{H} &= j\omega\varepsilon_0(\nabla\varepsilon_r \times \mathbf{E} + \varepsilon_r \nabla \times \mathbf{E}) \\ &= j\omega\varepsilon_0(\nabla\varepsilon_r \times \mathbf{E}) + j\omega\varepsilon_0\varepsilon_r(-j\omega\mu_0 \mathbf{H}) \\ &= j\omega\varepsilon_0(\nabla\varepsilon_r \times \mathbf{E}) + k_0^2 \varepsilon_r \mathbf{H}. \end{aligned} \tag{1.34}$$

Using

$$\mathbf{E} = \frac{1}{j\omega\varepsilon_0\varepsilon_r} \nabla \times \mathbf{H} \tag{1.35}$$

obtained from Eqs. (1.19) and (1.20), we get from Eq. (1.30) the following vectorial wave equation for the magnetic field **H**:

$$\nabla^2 \mathbf{H} + \frac{\nabla \varepsilon_r}{\varepsilon_r} \times (\nabla \times \mathbf{H}) + k_0^2 \varepsilon_r \mathbf{H} = \mathbf{0}. \tag{1.36}$$

Using Eq. (1.31), we can rewrite Eq. (1.36) as

$$\nabla^2 \mathbf{H} + \frac{\nabla \varepsilon_r}{\varepsilon_r} \times (\nabla \times \mathbf{H}) + k^2 \mathbf{H} = \mathbf{0}. \tag{1.37}$$

When the relative permittivity ε_r is constant in the medium, this vectorial wave equation can be reduced to the Helmholtz equation

$$\nabla^2 \mathbf{H} + k^2 \mathbf{H} = \mathbf{0}. \tag{1.38}$$

Now, we discuss an optical waveguide whose structure is uniform in the z direction. The derivative of an electromagnetic field with respect to the z coordinate is constant such that

$$\frac{\partial}{\partial z} = -j\beta, \tag{1.39}$$

where β is the propagation constant and is the z-directed component of the wave number k. The ratio of the propagation constant in the medium, β, to the wave number in a vacuum, k_0, is called the effective index:

$$n_{\text{eff}} = \frac{\beta}{k_0}. \tag{1.40}$$

When λ_0 is the wavelength in a vacuum,

$$\beta = \frac{2\pi}{\lambda_0} n_{\text{eff}} = \frac{2\pi}{\lambda_0/n_{\text{eff}}} = \frac{2\pi}{\lambda_{\text{eff}}}, \tag{1.41}$$

where $\lambda_{\text{eff}} = \lambda_0/n_{\text{eff}}$ is the z-directed component of the wavelength in the medium. The physical meaning of the propagation constant β is the phase rotation per unit propagation distance. Thus, the effective index n_{eff} can be interpreted as the ratio of a wavelength in the medium to the wavelength in a vacuum, or as the ratio of a phase rotation in the medium to the phase rotation in a vacuum.

We can summarize the Helmholtz equation for the electric field **E** as

$$\nabla_\perp^2 \mathbf{E} + (k^2 - \beta^2)\mathbf{E} = \mathbf{0} \tag{1.42}$$

or

$$\nabla_\perp^2 \mathbf{E} + k_0^2(\varepsilon_r - n_{\text{eff}}^2)\mathbf{E} = \mathbf{0}. \tag{1.43}$$

For the magnetic field **H**, on the other hand, we get the Helmholtz equation

$$\nabla_\perp^2 \mathbf{H} + (k^2 - \beta^2)\mathbf{H} = \mathbf{0} \tag{1.44}$$

or

$$\nabla_\perp^2 \mathbf{H} + k_0^2(\varepsilon_r - n_{\text{eff}}^2)\mathbf{H} = \mathbf{0}, \tag{1.45}$$

where we used the definition $\nabla_\perp^2 = \partial^2/\partial x^2 + \partial^2/\partial y^2$.

1.3 POYNTING VECTORS

In this section, the time-dependent electric and magnetic fields are expressed as $\mathbf{E}(\mathbf{r}, t)$ and $\mathbf{H}(\mathbf{r}, t)$, and the time-independent electric and magnetic fields are expressed as $\bar{\mathbf{E}}(\mathbf{r})$ and $\bar{\mathbf{H}}(\mathbf{r})$. Because the voltage is the integral of an electric field and because the magnetic field is created by a current, the product of the electric field and the magnetic field is related to the energy of the electromagnetic fields. Applying a divergence operator $\nabla \cdot$ to $\mathbf{E} \times \mathbf{H}$, we get

$$\nabla \cdot (\mathbf{E} \times \mathbf{H}) = \mathbf{H} \cdot \nabla \times \mathbf{E} + \mathbf{E} \cdot \nabla \times \mathbf{H}.$$

Substituting Maxwell's equations (1.8) and (1.9) into this equation, we get

$$\begin{aligned}\nabla \cdot (\mathbf{E} \times \mathbf{H}) &= -\mu \mathbf{H} \cdot \frac{\partial \mathbf{H}}{\partial t} - \varepsilon \mathbf{E} \cdot \frac{\partial \mathbf{H}}{\partial t} - \sigma \mathbf{E}^2 \\ &= -\frac{\partial}{\partial t}\left(\frac{1}{2}\varepsilon \mathbf{E}^2 + \frac{1}{2}\mu \mathbf{H}^2\right) - \sigma \mathbf{E}^2.\end{aligned} \tag{1.46}$$

When Eq. (1.46) is integrated over a volume V, we get

$$\int_V \nabla \cdot (\mathbf{E} \times \mathbf{H}) \, dV = \int_S (\mathbf{E} \times \mathbf{H})_n \, dS$$
$$= -\frac{\partial}{\partial t} \int_V \left(\frac{1}{2}\varepsilon \mathbf{E}^2 + \frac{1}{2}\mu \mathbf{H}^2\right) dV - \int_V \sigma \mathbf{E}^2 \, dV, \tag{1.47}$$

where we make use of Gauss's law and n designates a component normal to the surface S of the volume V.

The first two terms of the last equation correspond to the rate of the reduction of the stored energy in volume V per unit time, while the third term corresponds to the rate of reduction of the energy due to Joule heating in volume V per unit time. Thus, the term $\int_S (\mathbf{E} \times \mathbf{H})_n \, dS$ is considered to be the rate of energy loss through the surface.

Thus,

$$\mathbf{S} = \mathbf{E} \times \mathbf{H} \tag{1.48}$$

is the energy that passes through a unit area per unit time. It is called a Poynting vector.

For an electromagnetic wave that oscillates at a single angular frequency ω, the time-averaged Poynting vector $\langle \mathbf{S} \rangle$ is calculated as follows:

$$\begin{aligned}
\langle \mathbf{S} \rangle &= \langle \mathbf{E} \times \mathbf{H} \rangle \\
&= \langle \text{Re}\{\bar{\mathbf{E}}(\mathbf{r}) \exp(j\omega t)\} \times \text{Re}\{\bar{\mathbf{H}}(\mathbf{r}) \exp(j\omega t)\} \rangle \\
&= \left\langle \frac{\bar{\mathbf{E}} \exp(j\omega t) + \bar{\mathbf{E}}^* \exp(-j\omega t)}{2} \times \frac{\bar{\mathbf{H}} \exp(j\omega t) + \bar{\mathbf{H}}^* \exp(-j\omega t)}{2} \right\rangle \\
&= \tfrac{1}{4} \langle (\bar{\mathbf{E}} \times \bar{\mathbf{H}}^* + \bar{\mathbf{E}}^* \times \bar{\mathbf{H}} + \bar{\mathbf{E}} \times \bar{\mathbf{H}} \exp(j2\omega t) + \bar{\mathbf{E}}^* \times \bar{\mathbf{H}}^* \exp(-j2\omega t)) \rangle \\
&= \tfrac{1}{2} \text{Re}\{\langle \bar{\mathbf{E}} \times \bar{\mathbf{H}}^* \rangle\}.
\end{aligned} \tag{1.49}$$

Here, we used $\langle \exp(j2\omega t) \rangle = \langle \exp(-j2\omega t) \rangle = 0$.

Thus, for an electromagnetic wave oscillating at a single angular frequency, the quantity

$$\mathbf{S} = \tfrac{1}{2} \bar{\mathbf{E}} \times \bar{\mathbf{H}}^* \tag{1.50}$$

is defined as a complex Poynting vector and the energy actually propagating is considered to be the real part of it.

1.4 BOUNDARY CONDITIONS FOR ELECTROMAGNETIC FIELDS

The boundary conditions required for the electromagnetic fields are summarized as follows:

(a) Tangential components of the electric fields are continuous such that

$$E_{1t} = E_{2t}. \tag{1.51}$$

(b) When no current flows on the surface, tangential components of the magnetic fields are continuous such that

$$H_{1t} = H_{2t}. \tag{1.52}$$

When a current flows on the surface, the magnetic fields are discontinuous and are related to the current density J_S as follows:

$$H_{1t} - H_{2t} = J_S. \tag{1.53}$$

Or, since the magnetic field and the current are perpendicular to each other, the vectorial representation is

$$\mathbf{n} \times (\mathbf{H}_1 - \mathbf{H}_2) = \mathbf{J}_S. \tag{1.54}$$

(c) When there is no charge on the surface, the normal components of the electric flux densities are continuous such that

$$D_{1n} = D_{2n}. \tag{1.55}$$

When there are charges on the surface, the electric flux densities are discontinuous and are related to the charge density ρ_S as follows:

$$D_{1n} - D_{2n} = \rho_S. \tag{1.56}$$

(d) Normal components of the magnetic flux densities are continuous such that

$$B_{1n} = B_{2n}. \tag{1.57}$$

Here, the vectors **n** and **t** in these equations are respectively unit normal and tangential vectors at the boundary.

PROBLEMS

1. Use Maxwell's equations to specify the features of a plane wave propagating in a homogeneous nonconductive medium.

ANSWER

Maxwell's equations are written as

$$\frac{\partial E_z}{\partial y} - \frac{\partial E_y}{\partial z} = -j\omega\mu_0 H_x, \tag{P1.1}$$

$$\frac{\partial E_x}{\partial z} - \frac{\partial E_z}{\partial x} = -j\omega\mu_0 H_y, \tag{P1.2}$$

$$\frac{\partial E_y}{\partial x} - \frac{\partial E_x}{\partial y} = -j\omega\mu_0 H_z, \tag{P1.3}$$

$$\frac{\partial H_z}{\partial y} - \frac{\partial H_y}{\partial z} = j\omega\varepsilon_0\varepsilon_r E_x, \tag{P1.4}$$

$$\frac{\partial H_x}{\partial z} - \frac{\partial H_z}{\partial x} = j\omega\varepsilon_0\varepsilon_r E_y, \tag{P1.5}$$

$$\frac{\partial H_y}{\partial x} - \frac{\partial H_x}{\partial y} = j\omega\varepsilon_0\varepsilon_r E_z. \tag{P1.6}$$

Since the electric and magnetic fields of the plane wave depend not on the x and y coordinates but on the z coordinate, the derivatives with respect to the coordinates for directions other than the propagation direction are zero. That is, $\partial/\partial x = \partial/\partial y = 0$.

From Eqs. (P1.3) and (P1.6), we get

$$H_z = E_z = 0, \quad \text{(P1.7)}$$

The remaining equations are

$$\frac{dE_y}{dz} = j\omega\mu_0 H_x, \quad \text{(P1.8)}$$

$$\frac{dE_x}{dz} = -j\omega\mu_0 H_y, \quad \text{(P1.9)}$$

$$\frac{dH_y}{dz} = -j\omega\varepsilon_0\varepsilon_r E_x, \quad \text{(P1.10)}$$

$$\frac{dH_x}{dz} = j\omega\varepsilon_0\varepsilon_r E_y. \quad \text{(P1.11)}$$

Equations (P1.8)–(P1.11) are categorized into two sets:

Set 1: $\quad \dfrac{dE_x}{dz} = -j\omega\mu_0 H_y \quad$ and $\quad \dfrac{dH_y}{dz} = -j\omega\varepsilon_0\varepsilon_r E_x.$ (P1.12)

Set 2: $\quad \dfrac{dE_y}{dz} = j\omega\mu_0 H_x \quad$ and $\quad \dfrac{dH_x}{dz} = j\omega\varepsilon_0\varepsilon_r E_y.$ (P1.13)

The equations of set 1 can be reduced to

$$\frac{d^2 E_x}{dz^2} + k^2 E_x = 0 \quad \text{and} \quad \frac{d^2 H_y}{dz^2} + k^2 H_y = 0, \quad \text{(P1.14)}$$

where $k^2 = \omega^2 \varepsilon_0 \mu_0 \varepsilon_r = k_0^2 \varepsilon_r$. And the equations of set 2 can be reduced to

$$\frac{d^2 E_y}{dz^2} + k^2 E_y = 0 \quad \text{and} \quad \frac{d^2 H_x}{dz^2} + k^2 H_x = 0. \quad \text{(P1.15)}$$

Here, we discuss a plane wave propagating in the z direction. Considering that Eq. (P1.14) implies that both the electric field component E_x and the magnetic field component H_y propagate with the wave number k, where it should be noted that the propagation constant β is equal to the wave number k in this case and that the pure imaginary number j $[= \exp(\frac{1}{2}j\pi)]$ corresponds to phase rotation by $90°$, we can illustrate the propagation of the electric field component E_x and the magnetic field

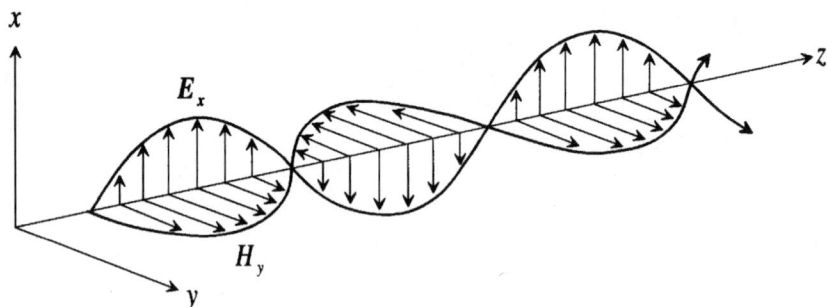

FIGURE P1.1. Propagation of an electromagnetic field.

component H_y, as shown in Fig. P1.1. When we substitute E_x for E_y and $-H_y$ for H_x, the equations of set 2 are equivalent to those of set 1. Since the field components in set 2 can be obtained by rotating the field components in set 1 by 90°, sets 1 and 2 are basically equivalent to each other.

The features of the plane wave are summarized as follows: (1) the electric and magnetic fields are uniform in directions perpendicular to the propagation direction, that is, $\partial/\partial x = \partial/\partial y = 0$; (2) the fields have no component in the propagation direction, that is, $H_z = E_z = 0$; (3) the electric field and the magnetic field components are perpendicular to each other; and (4) the propagation direction is the direction in which a screw being turned to the right, as if the electric field component were being turned toward the magnetic field component, advances.

2. Under the assumption that the relative permeability in the medium is equal to 1 and that a plane wave propagates in the $+z$ direction, prove that $\sqrt{\mu_0} H_y = \sqrt{\varepsilon} E_x$.

ANSWER

The derivative with respect to the z coordinate can be reduced to $d/dz = -jk = -j\omega\sqrt{\varepsilon\mu_0}$ by using Eq. (P1.14). Thus, the relation follows from Eq. (P1.12).

REFERENCE

[1] R. E. Collin, *Foundations for Microwave Engineering*, McGraw-Hill, New York, 1966.

CHAPTER 2

ANALYTICAL METHODS

Before discussing the numerical methods in Chapters 3–7, we first describe analytical methods: a method for a three-layer slab optical waveguide, an effective index method, and Marcatili's method. For actual optical waveguides, the analytical methods are less accurate than the numerical methods, but they are easier to use and more transparent. In this chapter, we also discuss a cylindrical coordinate analysis of the step-index optical fiber.

2.1 METHOD FOR A THREE-LAYER SLAB OPTICAL WAVEGUIDE

In this section, we discuss an analysis for a three-layer slab optical waveguide with a one-dimensional (1D) structure. The reader is referred to the literature for analyses of other multilayer structures [1, 2].

Figure 2.1 shows a three-layer slab optical waveguide with refractive indexes n_1, n_2, and n_3. Its structure is uniform in the y and z directions. Regions 1 and 3 are cladding layers, and region 2 is a core layer that has a refractive index higher than that of the cladding layers. Since the tangential field components are connected at the interfaces between adjacent media, we can start with the Helmholtz equations (1.47) and (1.49), which are for uniform media. Furthermore, since the structure is

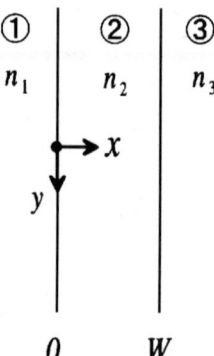

FIGURE 2.1. Three-layer slab optical waveguide.

uniform in the y direction, we can assume $\partial/\partial y = 0$. Thus, the equation for the electric field **E** is

$$\frac{d^2\mathbf{E}}{dx^2} + k_0^2(\varepsilon_r - n_{\text{eff}}^2)\mathbf{E} = \mathbf{0}. \tag{2.1}$$

Similarly, we easily get the equation for the magnetic field **H**:

$$\frac{d^2\mathbf{H}}{dx^2} + k_0^2(\varepsilon_r - n_{\text{eff}}^2)\mathbf{H} = \mathbf{0}. \tag{2.2}$$

Next, we discuss the two modes that propagate in the three-layer slab optical waveguide: the transverse electric mode (TE mode) and the transverse magnetic mode (TM mode). For better understanding, we again derive the wave equation from Maxwell's equations

$$\nabla \times \mathbf{E} = -j\omega\mu_0 \mathbf{H}, \tag{2.3}$$

$$\nabla \times \mathbf{H} = j\omega\varepsilon_0\varepsilon_r \mathbf{E}, \tag{2.4}$$

whose component representations are

$$\frac{\partial E_z}{\partial y} - \frac{\partial E_y}{\partial z} = -j\omega\mu_0 H_x, \tag{2.5}$$

$$\frac{\partial E_x}{\partial z} - \frac{\partial E_z}{\partial x} = -j\omega\mu_0 H_y, \tag{2.6}$$

$$\frac{\partial E_y}{\partial x} - \frac{\partial E_x}{\partial y} = -j\omega\mu_0 H_z, \tag{2.7}$$

2.1 METHOD FOR A THREE-LAYER SLAB OPTICAL WAVEGUIDE

$$\frac{\partial H_z}{\partial y} - \frac{\partial H_y}{\partial z} = j\omega\varepsilon_0\varepsilon_r E_x, \tag{2.8}$$

$$\frac{\partial H_x}{\partial z} - \frac{\partial H_z}{\partial x} = j\omega\varepsilon_0\varepsilon_r E_y, \tag{2.9}$$

$$\frac{\partial H_y}{\partial x} - \frac{\partial H_x}{\partial y} = j\omega\varepsilon_0\varepsilon_r E_z. \tag{2.10}$$

As mentioned in Chapter 1, we assume here that the relative permeability $\mu_r = 1$. That is, $\mu = \mu_r\mu_0 = \mu_0$. Since the structure is uniform in the propagation direction, the derivative with respect to the z coordinate, $\partial/\partial z$, can be replaced by $-j\beta$. The effective index can be expressed as $n_{\text{eff}} = \beta/k_0$, where k_0 is the wave number in a vacuum.

2.1.1 TE Mode

In the TE mode, the electric field is not in the longitudinal direction ($E_z = 0$) but in the transverse direction ($E_y \neq 0$). Since the structure is uniform in the y direction, $\partial/\partial y = 0$. Substitution of these relations into Eq. (2.10) results in $\partial H_y/\partial x = 0$. Since this means that H_y is constant, we can assume that $H_y = 0$. Furthermore, substitution of $E_z = H_y = 0$ into Eq. (2.6) results in $\partial E_x/\partial z = 0$, which means that $E_x = 0$. We thus get

$$E_x = E_z = H_y = 0. \tag{2.11}$$

Substituting $H_x = -(\beta/\omega\mu_0)E_y$, derived from Eq. (2.5), and $H_z = (j/\omega\mu_0)\,\partial E_y/\partial x$, derived from Eq. (2.7), into Eq. (2.9), we get the following wave equation for the principal electric field component E_y:

$$\frac{d^2 E_y}{dx^2} + k_0^2(\varepsilon_r - n_{\text{eff}}^2)E_y = 0, \tag{2.12}$$

where $k_0 = \omega\sqrt{\varepsilon_0\mu_0}$.

Next, we derive the characteristic equation used to calculate the effective index n_{eff}. The principal electric field component E_y in regions 1, 2, and 3 can be expressed as

$$E_y(x) = C_1 \exp(\gamma_1 x) \quad \text{as} \quad \gamma_1 = k_0\sqrt{n_{\text{eff}}^2 - n_1^2} \quad \text{(region 1)}, \tag{2.13}$$

$$= C_2 \cos(\gamma_2 x + \alpha) \quad \text{as} \quad \gamma_2 = k_0\sqrt{n_2^2 - n_{\text{eff}}^2} \quad \text{(region 2)}, \tag{2.14}$$

$$= C_3 \exp[-\gamma_3(x - W)] \quad \text{as} \quad \gamma_3 = k_0\sqrt{n_{\text{eff}}^2 - n_3^2} \quad \text{(region 3)}. \tag{2.15}$$

Here, C_1, C_2, and C_3 are unknown constants. Since the number of unknowns is 4 (n_{eff}, C_1, C_2, and C_3), four equations are needed to determine the effective index n_{eff}. To obtain the four equations, we impose boundary conditions on the tangential electric field component E_y and the tangential magnetic field component H_z at $x = 0$ and $x = W$. The tangential magnetic field component H_z is

$$H_z = \frac{-1}{j\omega\mu_0} \frac{\partial E_y}{\partial x}, \tag{2.16}$$

which for the three regions is expressed as

$$H_z(x) = -\frac{\gamma_1}{j\omega\mu_0} C_1 \exp(\gamma_1 x) \quad \text{(region 1)}, \tag{2.17}$$

$$= \frac{\gamma_2}{j\omega\mu_0} C_2 \sin(\gamma_2 x + \alpha) \quad \text{(region 2)}, \tag{2.18}$$

$$= \frac{\gamma_3}{j\omega\mu_0} C_3 \exp[-\gamma_3(x - W)] \quad \text{(region 3)}. \tag{2.19}$$

Since the boundary condition requirement is that the *tangential electric field components as well as the tangential magnetic field components are equal at the interfaces between adjacent media*, the boundary conditions on these field components at $x = 0$ are expressed as

$$E_{y1}(0) = E_{y2}(0), \tag{2.20}$$

$$H_{z1}(0) = H_{z2}(0) \tag{2.21}$$

2.1 METHOD FOR A THREE-LAYER SLAB OPTICAL WAVEGUIDE

and at $x = W$ are expressed as

$$E_{y2}(W) = E_{y3}(W), \tag{2.22}$$

$$H_{z2}(W) = H_{z3}(W). \tag{2.23}$$

The resultant equations are

$$C_1 = C_2 \cos\alpha \quad \text{[from Eq. (2.20)]}, \tag{2.24}$$

$$-\gamma_1 C_1 = \gamma_2 C_2 \sin\alpha \quad \text{[from Eq. (2.21)]}, \tag{2.25}$$

$$C_2 \cos(\gamma_2 W + \alpha) = C_3 \quad \text{[from Eq. (2.22)]}, \tag{2.26}$$

$$-\gamma_2 C_2 \sin(\gamma_2 W + \alpha) = -\gamma_3 C_3 \quad \text{[from Eq. (2.23)]}. \tag{2.27}$$

Thus, dividing Eq. (2.25) by Eq. (2.24), we get

$$\alpha = -\tan^{-1}\left(\frac{\gamma_1}{\gamma_2}\right) + q_1\pi \quad (q_1 = 0, 1, 2, \ldots). \tag{2.28}$$

On the other hand, dividing Eq. (2.27) by Eq. (2.26), we get

$$\gamma_2 W = \tan^{-1}\left(\frac{\gamma_3}{\gamma_2}\right) - \alpha + q_2\pi \quad (q_2 = 0, 1, 2, \ldots). \tag{2.29}$$

Substitution of α in Eq. (2.28) into Eq. (2.29) results in the following characteristic equation:

$$\gamma_2 W = \tan^{-1}\left(\frac{\gamma_1}{\gamma_2}\right) + \tan^{-1}\left(\frac{\gamma_3}{\gamma_2}\right) + q\pi \quad (q = 0, 1, 2, \ldots). \tag{2.30}$$

Or, using

$$\tan^{-1}\left(\frac{y}{x}\right) = \frac{\pi}{2} - \tan^{-1}\left(\frac{x}{y}\right), \tag{2.31}$$

we can rewrite this equation as

$$\gamma_2 W = -\tan^{-1}\left(\frac{\gamma_2}{\gamma_1}\right) - \tan^{-1}\left(\frac{\gamma_2}{\gamma_3}\right) + (q+1)\pi \quad (q = 0, 1, 2, \ldots). \tag{2.32}$$

2.1.2 TM Mode

In the TM mode, the magnetic field component is not in the longitudinal direction ($H_z = 0$) but in the transverse direction ($H_y \neq 0$). Since the structure is uniform in the y direction, $\partial/\partial y = 0$. Thus, we get $\partial E_y/\partial x = 0$ from Eq. (2.7). Since this means that E_y is constant, we can assume that $E_y = 0$. Furthermore, substitution of $H_z = E_y = 0$ into Eq. (2.9) results in $\partial H_x/\partial z = 0$, which means that $H_x = 0$. We thus get

$$H_x = H_z = E_y = 0. \tag{2.33}$$

Substituting $E_x = (\beta/\omega\varepsilon_0\varepsilon_r)H_y$, derived from Eq. (2.8), and $E_z = -(j/\omega\varepsilon_0\varepsilon_r)\partial H_y/\partial x$, derived from Eq. (2.10), into Eq. (2.6), we get the following wave equation for the principal magnetic field component H_y:

$$\frac{d^2 H_y}{dx^2} + k_0^2(\varepsilon_r - n_{\text{eff}}^2)H_y = 0. \tag{2.34}$$

The field components on which the boundary conditions should be imposed are the principal magnetic field component H_y and the longitudinal electric field component E_z. The principal magnetic field component H_y can be expressed as

$$H_y(x) = C_1 \exp(\gamma_1 x) \quad \text{as} \quad \gamma_1 = k_0\sqrt{n_{\text{eff}}^2 - n_1^2} \quad \text{(region 1)}, \tag{2.35}$$

$$= C_2 \cos(\gamma_2 x + \alpha) \quad \text{as} \quad \gamma_2 = k_0\sqrt{n_2^2 - n_{\text{eff}}^2} \quad \text{(region 2)}, \tag{2.36}$$

$$= C_3 \exp[-\gamma_3(x - W)] \quad \text{as} \quad \gamma_3 = k_0\sqrt{n_{\text{eff}}^2 - n_3^2} \quad \text{(region 3)}. \tag{2.37}$$

2.1 METHOD FOR A THREE-LAYER SLAB OPTICAL WAVEGUIDE

The tangential electric field component E_z is

$$E_z = \frac{1}{j\omega\varepsilon_0\varepsilon_r} \frac{\partial H_y}{\partial x}, \qquad (2.38)$$

which is expressed as

$$E_z(x) = \frac{\gamma_1}{j\omega\varepsilon_0\varepsilon_r} C_1 \exp(\gamma_1 x) \qquad \text{(region 1)}, \qquad (2.39)$$

$$= -\frac{\gamma_2}{j\omega\varepsilon_0\varepsilon_r} C_2 \sin(\gamma_2 x + \alpha) \qquad \text{(region 2)}, \qquad (2.40)$$

$$= \frac{\gamma_3}{j\omega\varepsilon_0\varepsilon_r} C_3 \exp[-\gamma_3(x - W)] \qquad \text{(region 3)}. \qquad (2.41)$$

Imposing the boundary conditions on the tangential fields at $x = 0$ and $x = W$, we get

$$C_1 = C_2 \cos\alpha \qquad [\text{from } H_{y1}(0) = H_{y2}(0)], \qquad (2.42)$$

$$-\frac{\gamma_1}{\varepsilon_{r1}} C_1 = \frac{\gamma_1}{\varepsilon_{r2}} C_2 \sin\alpha \qquad [\text{from } E_{z1}(0) = E_{z2}(0)], \qquad (2.43)$$

$$C_2 \cos(\gamma_2 W + \alpha) = C_3 \qquad [\text{from } H_{y2}(W) = H_{y3}(W)], \qquad (2.44)$$

$$-\frac{\gamma_2}{\varepsilon_{r2}} C_2 \sin(\gamma_2 W + \alpha) = -\frac{\gamma_2}{\varepsilon_{r2}} C_3 \qquad [\text{from } E_{z2}(W) = E_{z3}(W)]. \qquad (2.45)$$

Dividing Eq. (2.43) by Eq. (2.42), we get

$$\alpha = -\tan^{-1}\left(\frac{\varepsilon_{r2}}{\varepsilon_{r1}} \frac{\gamma_1}{\gamma_2}\right) + q_1\pi \qquad (q_1 = 0, 1, 2, \ldots). \qquad (2.46)$$

On the other hand, dividing Eq. (2.45) by Eq. (2.44), we get

$$\gamma_2 W = \tan^{-1}\left(\frac{\varepsilon_{r2}}{\varepsilon_{r3}} \frac{\gamma_3}{\gamma_2}\right) - \alpha + q_2\pi \qquad (q_2 = 0, 1, 2, \ldots). \qquad (2.47)$$

Substitution of the variable α in Eq. (2.46) into Eq. (2.47) results in the following characteristic equation:

$$\gamma_2 W = \tan^{-1}\left(\frac{\varepsilon_{r2}}{\varepsilon_{r1}}\frac{\gamma_1}{\gamma_2}\right) + \tan^{-1}\left(\frac{\varepsilon_{r2}}{\varepsilon_{r3}}\frac{\gamma_3}{\gamma_2}\right) + q\pi \quad (q = 0, 1, 2, \ldots). \tag{2.48}$$

Using Eq. (2.31), we also get

$$\gamma_2 W = -\tan^{-1}\left(\frac{\varepsilon_{r1}}{\varepsilon_{r2}}\frac{\gamma_2}{\gamma_1}\right) - \tan^{-1}\left(\frac{\varepsilon_{r3}}{\varepsilon_{r2}}\frac{\gamma_2}{\gamma_3}\right) \\ + (q+1)\pi \quad (q = 0, 1, 2, \ldots). \tag{2.49}$$

Comparing the characteristic equations (2.30) and (2.32) for the TE mode and Eqs. (2.48) and (2.49) for the TM mode, one discovers that the characteristic equations for the TM mode contain the ratio of the relative permittivities of adjacent media.

2.2 EFFECTIVE INDEX METHOD

Here, we discuss the effective index method, which allows us to analyze two-dimensional (2D) optical waveguide structures by simply repeating the slab optical waveguide analyses.

Figure 2.2 shows an example of a 2D optical waveguide and illustrates the concept of the effective index method. We consider the scalar wave equation

$$\frac{\partial^2 \phi(x,y)}{\partial x^2} + \frac{\partial^2 \phi(x,y)}{\partial y^2} + k_0^2(\varepsilon_r(x,y) - n_{\text{eff}}^2)\phi(x,y) = 0, \tag{2.50}$$

where n_{eff} is the effective index to be obtained. We separate the wave function $\phi(x, y)$ into two functions:

$$\phi(x, y) = f(x) \cdot g(y). \tag{2.51}$$

2.2 EFFECTIVE INDEX METHOD

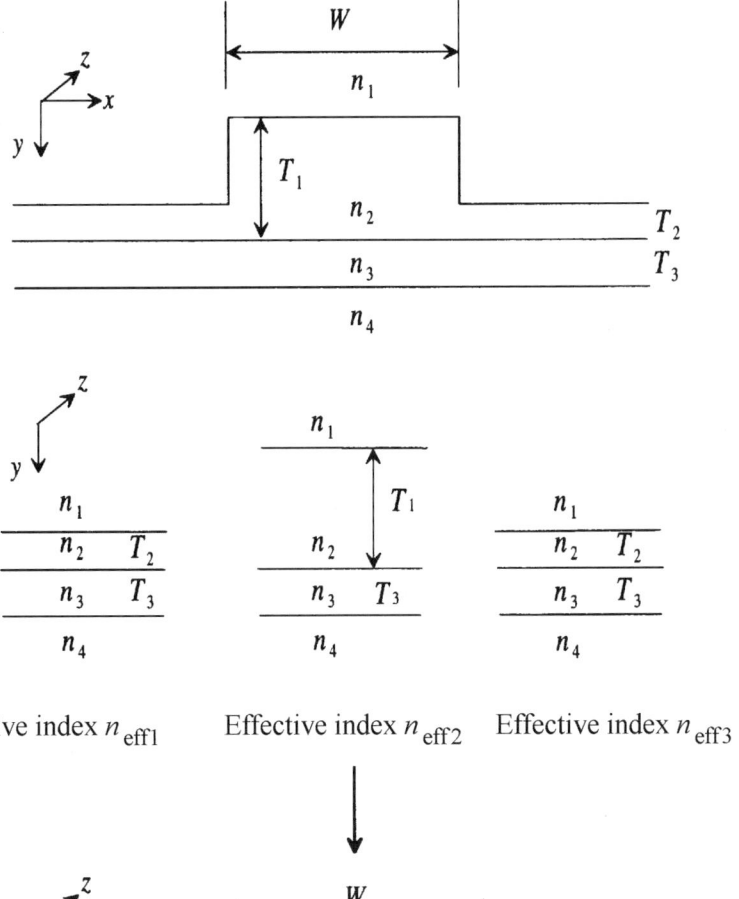

FIGURE 2.2. Concept of the effective index method.

This corresponds to the assumption that there is no interaction between the variables x and y. Substituting Eq. (2.51) into Eq. (2.50) and dividing the resultant equation by the wave function $\phi(x, y)$, we get

$$\frac{1}{f(x)}\frac{d^2 f(x)}{dx^2} + \frac{1}{g(y)}\frac{d^2 g(y)}{dy^2} + k_0^2(\varepsilon_r(x, y) - n_{\text{eff}}^2) = 0. \qquad (2.52)$$

Setting the sum of the second and third terms of Eq. (2.52) equal to $k_0^2 N^2(x)$, we get

$$\frac{1}{g(y)}\frac{d^2 g(y)}{dy^2} + k_0^2 \varepsilon_r(x, y) = k_0^2 N^2(x). \tag{2.53}$$

This means that the sum of the first and fourth terms is equal to $-k_0^2 N^2(x)$:

$$\frac{1}{f(x)}\frac{d^2 f(x)}{dx^2} - k_0^2 n_{\text{eff}}^2 = -k_0^2 N^2(x). \tag{2.54}$$

Through these procedures we get the two independent equations

$$\frac{d^2 g(y)}{dy^2} + k_0^2 [\varepsilon_r(x, y) - N^2(x)] g(y) = 0 \tag{2.55}$$

and

$$\frac{d^2 f(x)}{dx^2} + k_0^2 (N^2(x) - n_{\text{eff}}^2) f(x) = 0. \tag{2.56}$$

The effective index calculation procedure can be summarized as follows:

(a) As shown in Fig. 2.2, replace the 2D optical waveguide with a combination of 1D optical waveguides.
(b) For each 1D optical waveguide, calculate the effective index along the y axis.
(c) Model an optical slab waveguide by placing the effective indexes calculated in step (b) along the x axis.
(d) Obtain the effective index by solving the model obtained in step (c) along the x axis.

It should be noted that, for the TE mode of the 2D optical waveguide, we first do the TE-mode analysis and then the TM-mode analysis. And, for the TM-mode analysis of the 2D optical waveguide, we do these analyses in the opposite order.

2.3 MARCATILI'S METHOD

Here, we discuss Marcatili's method for analyzing 2D optical waveguides [3]. Proposed in 1969, it is still in wide use.

Figure 2.3 shows a cross-sectional view of a buried optical waveguide. The core has a refractive index n_1, width $2a$, and height $2b$. It is surrounded by cladding that has a refractive index n_2. In Marcatili's method, it is assumed that the electric fields and magnetic fields are confined to the core and do not exist in the four hatched regions shown in Fig. 2.3. Thus, the continuity conditions for the electric fields and the magnetic fields are imposed only at the interfaces of regions of 1 and 2, 1 and 3, 1 and 4, and 1 and 5.

We discuss the E_{pq}^x mode, which has E_x and H_y as principal field components, and the E_{pq}^y mode, which has E_y and H_x as principal field components. Here, p and q and are integers and respectively correspond to the numbers of peaks of optical power in the x and y directions. Thus, unlike ordinary mode orders, which begin from 0, they begin from 1.

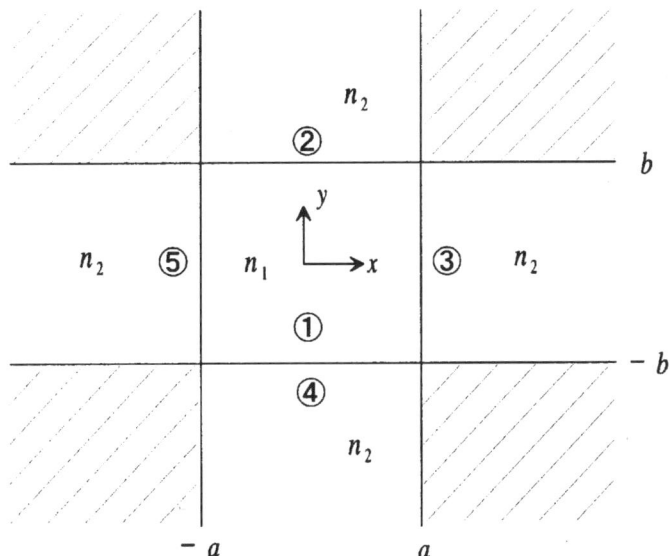

FIGURE 2.3. Marcatili's method.

2.3.1 E_{pq}^x Mode

The electric field of the E_{pq}^x mode is assumed to be polarized in the x direction, which results in $E_y = 0$. Since the structure of the optical waveguide is assumed to be invariant in the z direction, the derivative with respect to z is replaced by $-j\beta$. The component representations shown in Eqs. (2.5)–(2.10) are reduced to

$$\frac{\partial E_z}{\partial y} = -j\omega\mu_0 H_x, \tag{2.57}$$

$$-j\beta E_x - \frac{\partial E_z}{\partial x} = -j\omega\mu_0 H_y, \tag{2.58}$$

$$-\frac{\partial E_x}{\partial y} = -j\omega\mu_0 H_z, \tag{2.59}$$

$$\frac{\partial H_z}{\partial y} + j\beta H_y = j\omega\varepsilon_0\varepsilon_r E_x, \tag{2.60}$$

$$-j\beta H_x - \frac{\partial H_z}{\partial x} = 0. \tag{2.61}$$

Thus, we get

$$\frac{\partial H_y}{\partial x} - \frac{\partial H_x}{\partial y} = j\omega\varepsilon_0\varepsilon_r E_z, \tag{2.62}$$

where

$$H_x = \frac{-1}{j\omega\mu_0} \frac{\partial E_z}{\partial y} \qquad \text{[from Eq. (2.57)]}, \tag{2.63}$$

$$H_y = \frac{-1}{j\omega\mu_0}\left(-j\beta E_x - \frac{\partial E_z}{\partial x}\right) \qquad \text{[from Eq. (2.58)]}, \tag{2.64}$$

$$H_z = \frac{1}{j\omega\mu_0} \frac{\partial E_x}{\partial y} \qquad \text{[from Eq. (2.59)]}. \tag{2.65}$$

On the other hand, the component representation of the divergence equation

$$\nabla \cdot (\varepsilon_r \mathbf{E}) = 0 \tag{2.66}$$

is

$$\frac{\partial}{\partial x}(\varepsilon_r E_x) + (-j\beta)(\varepsilon_r E_z) = 0. \tag{2.67}$$

Thus, we get the longitudinal electric field

$$\begin{aligned} E_z &= \frac{1}{j\beta \varepsilon_r} \frac{\partial}{\partial x}(\varepsilon_r E_x) \\ &= \frac{1}{j\beta \varepsilon_r} \frac{\partial \varepsilon_r}{\partial x} E_x + \frac{1}{j\beta} \frac{\partial E_x}{\partial x} \\ &= \frac{1}{j\beta} \frac{\partial E_x}{\partial x}. \end{aligned} \tag{2.68}$$

Here, we assumed that $1/\varepsilon_r \cdot \partial \varepsilon_r/\partial x = 0$. That is, we ignored the dependence of the relative permittivity ε_r on the coordinate x in each region. This assumption will also be used for each element in the finite-element method discussed in Chapter 3. Eliminating E_z by substituting Eq. (2.68) into Eqs. (2.63) and (2.64), we get

$$H_x = \frac{-1}{j\omega\mu_0} \frac{\partial E_z}{\partial y} = \frac{1}{\omega\mu_0 \beta} \frac{\partial^2 E_x}{\partial x \, \partial y} \quad \text{[from Eq. (2.63)]} \tag{2.69}$$

$$\begin{aligned} H_y &= \frac{-1}{j\omega\mu_0}\left[-j\beta E_x - \frac{\partial}{\partial x}\left(\frac{1}{j\beta}\frac{\partial E_x}{\partial x}\right)\right] \\ &= \frac{1}{\omega\mu_0 \beta}\left(\beta^2 E_x - \frac{\partial^2 E_x}{\partial x^2}\right) \quad \text{[from Eq. (2.64)]}. \end{aligned} \tag{2.70}$$

The above results can be summarized as

$$H_x = \frac{1}{\omega\mu_0 \beta} \frac{\partial^2 E_x}{\partial x \, \partial y} \quad \text{[from Eq. (2.69)]}, \tag{2.71}$$

$$H_y = \frac{1}{\omega\mu_0 \beta}\left(\beta^2 E_x - \frac{\partial^2 E_x}{\partial x^2}\right) \quad \text{[from Eq. (2.70)]}, \tag{2.72}$$

$$H_z = \frac{1}{j\omega\mu_0} \frac{\partial E_x}{\partial y} \quad \text{[from Eq. (2.65)]}, \tag{2.73}$$

$$E_z = \frac{1}{j\beta} \frac{\partial E_x}{\partial x} \quad \text{[from Eq. (2.68)]}. \tag{2.74}$$

Substituting Eqs. (2.72) and (2.73) into Eq. (2.60), we get a wave equation for a principal field component E_x for the E_{pq}^x mode such that

$$\frac{\partial^2 E_x}{\partial x^2} + \frac{\partial^2 E_x}{\partial y^2} + k_0^2(\varepsilon_r - n_{\text{eff}}^2)E_x = 0. \tag{2.75}$$

The electric field and the magnetic field components to be connected are E_x and H_z at the boundaries $y = \pm b$ and are E_z and H_y at the boundaries $x = \pm a$.

Since the principal field component E_x is a solution (i.e., wave function) of the wave equation (2.75), we get the following field components in regions 1–5:

$$E_x = C_1 \cos(k_x x + \alpha_1) \cos(k_y y + \alpha_2) \quad \text{(region 1)}, \tag{2.76}$$

$$= C_2 \cos(k_x x + \alpha_1) \exp(-\gamma_y(y - b)) \quad \text{(region 2)}, \tag{2.77}$$

$$= C_3 \exp[-\gamma_x(x - a)] \cos(k_y y + \alpha_2) \quad \text{(region 3)}, \tag{2.78}$$

$$= C_4 \cos(k_x x + \alpha_1) \exp[\gamma_y(y + b)] \quad \text{(region 4)}, \tag{2.79}$$

$$= C_5 \exp[\gamma_x(x + a)] \cos(k_y y + \alpha_2) \quad \text{(region 5)}. \tag{2.80}$$

Substituting these wave functions into the wave equation (2.75), we get (after some mathematical manipulations) the following relations for the wave numbers:

$$k_x^2 + k_y^2 + \beta^2 = k_0^2 n_1^2, \tag{2.81}$$

$$k_x^2 - \gamma_y^2 + \beta^2 = k_0^2 n_2^2, \tag{2.82}$$

$$-\gamma_x^2 + k_y^2 + \beta^2 = k_0^2 n_2^2. \tag{2.83}$$

Subtracting Eq. (2.81) from Eq. (2.83) and Eq. (2.81) from Eq. (2.82), we get

$$\gamma_x^2 = k_0^2(n_1^2 - n_2^2) - k_x^2, \tag{2.84}$$

$$\gamma_y^2 = k_0^2(n_1^2 - n_2^2) - k_y^2. \tag{2.85}$$

The next step is to impose the boundary conditions specified by Eqs. (1.55) and (1.56) on the electric and magnetic fields.

A. Connection at $y = \pm b$**:** E_x **and** H_z Setting the electric field components E_x of regions 1 and 2 equal at $y = b$, from Eqs. (2.76) and (2.77), we get

$$C_1 \cos(k_y b + \alpha_2) = C_2. \tag{2.86}$$

And setting the magnetic field components H_z of regions 1 and 2, obtained by substituting Eqs. (2.76) and (2.77) into Eq. (2.73), equal at $y = b$, we get

$$C_1 k_y \sin(k_y b + \alpha_2) = C_2 \gamma_y. \tag{2.87}$$

Dividing Eq. (2.87) by Eq. (2.86), we get

$$\tan(k_y b + \alpha_2) = \frac{\gamma_y}{k_y}.$$

Therefore

$$k_y b + \alpha_2 = \tan^{-1}\left(\frac{\gamma_y}{k_y}\right) + q_1 \pi \quad (q_1 = 0, 1, \ldots). \tag{2.88}$$

On the other hand, setting the electric field components E_x of regions 1 and 4 equal at $y = -b$, from Eqs. (2.76) and (2.79), we get

$$C_1 \cos(-k_y b + \alpha_2) = C_4. \tag{2.89}$$

And setting the magnetic field components H_z of regions 1 and 4, obtained by substituting Eqs. (2.76) and (2.79) into Eq. (2.73), equal at $y = -b$, we get

$$C_1 k_y \sin(k_y b - \alpha_2) = C_4 \gamma_y. \tag{2.90}$$

Dividing Eq. (2.90) by Eq. (2.89), we get

$$k_y b - \alpha_2 = \tan^{-1}\left(\frac{\gamma_y}{k_y}\right) + q_2 \pi \quad (q_2 = 0, 1, \ldots). \tag{2.91}$$

And adding Eq. (2.88) to Eq. (2.91), we get

$$k_y b = \tan^{-1}\left(\frac{\gamma_y}{k_y}\right) + \tfrac{1}{2}(q - 1)\pi \quad (q = 1, 2, \ldots). \tag{2.92}$$

B. Connection at $x = \pm a$**:** E_z **and** H_y Setting the electric field components E_z of regions 1 and 3, obtained by substituting Eqs. (2.76) and (2.78) into Eq. (2.74), equal at $x = a$, we get

$$C_1 k_x \sin(k_x a + \alpha_1) = C_3 \gamma_x. \tag{2.93}$$

And setting the magnetic field components H_y of regions 1 and 3, obtained by substituting Eqs. (2.76) and (2.78) into Eq. (2.72), equal at $x = a$, we get

$$(\beta^2 + k_x^2) C_1 \cos(k_x a + \alpha_1) = (\beta^2 - \gamma_x^2) C_3. \tag{2.94}$$

Dividing Eq. (2.93) by Eq. (2.94), we get

$$\tan(k_x a + \alpha_1) = \frac{(\beta^2 + k_x^2)\gamma_x}{(\beta^2 - \gamma_x^2)k_x}. \tag{2.95}$$

Substituting the relations

$$\beta^2 + k_x^2 = k_0^2 n_1^2 - k_y^2, \tag{2.96}$$

$$\beta^2 - \gamma_x^2 = k_0^2 n_2^2 - k_y^2, \tag{2.97}$$

which are obtained from Eqs. (2.81) and (2.83), into Eq. (2.95), we get

$$\tan(k_x a + \alpha_1) = \frac{(k_0^2 n_1^2 - k_y^2)\gamma_x}{(k_0^2 n_2^2 - k_y^2)k_x}.$$

Therefore

$$k_x a + \alpha_1 = \tan^{-1}\left(\frac{(k_0^2 n_1^2 - k_y^2)\gamma_x}{(k_0^2 n_2^2 - k_y^2)k_x}\right) + p_1 \pi \qquad (p_1 = 0, 1, \ldots). \qquad (2.98)$$

On the other hand, setting the electric field components E_z of regions 1 and 5, obtained by substituting Eqs. (2.76) and (2.78) into Eq. (2.74), equal at $x = -a$, we get

$$C_1 k_x \sin(k_x a - \alpha_1) = C_5 \gamma_x. \qquad (2.99)$$

And setting the magnetic field components H_y of regions 1 and 5, obtained by substituting Eqs. (2.76) and (2.78) into Eq. (2.72), equal at $x = -a$, we get

$$(\beta^2 + k_x^2)C_1 \cos(-k_x a + \alpha_1) = (\beta^2 - \gamma_x^2)C_5. \qquad (2.100)$$

Substituting Eqs. (2.96) and (2.97) into the equation derived by dividing Eq. (2.99) by Eq. (2.100), we get

$$\tan(k_x a - \alpha_1) = \frac{(\beta^2 + k_x^2)\gamma_x}{(\beta^2 - \gamma_x^2)k_x}.$$

Therefore

$$\tan(k_x a - \alpha_1) = \frac{(k_0^2 n_1^2 - k_y^2)\gamma_x}{(k_0^2 n_2^2 - k_y^2)k_x},$$

and

$$k_x a - \alpha_1 = \tan^{-1}\left(\frac{(k_0^2 n_1^2 - k_y^2)\gamma_x}{(k_0^2 n_2^2 - k_y^2)k_x}\right) + p_2 \pi \quad (p_2 = 0, 1, \ldots). \quad (2.101)$$

Now, adding Eq. (2.98) to Eq. (2.101), we get

$$k_x a = \tan^{-1}\left(\frac{(k_0^2 \varepsilon_{r1} - k_y^2)}{(k_0^2 \varepsilon_{r2} - k_y^2)} \frac{\gamma_x}{k_x}\right) + \tfrac{1}{2}(p - 1)\pi \quad (p = 1, 2, \ldots),$$

(2.102)

where $\varepsilon_{r1} = n_1^2$ and $\varepsilon_{r2} = n_2^2$. Or, since $k_0 n_{1,2} \gg k_y$ for most cases, we get

$$k_x a = \tan^{-1}\left(\frac{\varepsilon_{r1}}{\varepsilon_{r2}} \frac{\gamma_x}{k_x}\right) + \tfrac{1}{2}(p - 1)\pi \quad (p = 1, 2, \ldots). \quad (2.103)$$

The propagation constant β (or the effective index n_{eff}) can be obtained as follows:

1. Obtain k_x by using a numerical technique such as a successive bisection method to solve Eq. (2.103) [or Eq. (2.102)]. (It should be noted that in each process of the successive bisection routine Eq. (2.84) can be used to obtain the γ_x corresponding to k_x.)
2. Similarly, obtain k_y by making use of Eqs. (2.85) and (2.92).
3. Finally, obtain the propagation constant β from Eq. (2.81).

Interesting points are as follows: Equation (2.92) corresponds to the characteristic equation of the TE mode for a three-layer slab optical waveguide parallel to the x axis. Equation (2.103), on the other hand, corresponds to the characteristic equation of the TM mode for a three-layer slab optical waveguide parallel to the y axis. The correspondence is the same as that described in the last part of Section 2.2. The x and y axes are related to each other through Eq. (2.81).

2.3.2 E_{pq}^{y} Mode

The magnetic field of the E_{pq}^{x} mode is assumed to be polarized in the x direction, which results in $H_y = 0$. The component representations shown in Eqs. (2.5)–(2.10) are reduced to

$$\frac{\partial H_z}{\partial y} = j\omega\varepsilon_0\varepsilon_r E_x, \tag{2.104}$$

$$-j\beta H_x - \frac{\partial H_z}{\partial x} = j\omega\varepsilon_0\varepsilon_r E_y, \tag{2.105}$$

$$-\frac{\partial H_x}{\partial y} = j\omega\varepsilon_0\varepsilon_r E_z, \tag{2.106}$$

$$\frac{\partial E_z}{\partial y} + j\beta E_y = -j\omega\mu_0 H_x, \tag{2.107}$$

$$-j\beta E_x - \frac{\partial E_z}{\partial x} = 0, \tag{2.108}$$

$$\frac{\partial E_y}{\partial x} - \frac{\partial E_x}{\partial y} = -j\omega\mu_0 H_z, \tag{2.109}$$

$$E_x = \frac{1}{j\omega\varepsilon_0\varepsilon_r} \frac{\partial H_z}{\partial y} \quad \text{[from Eq. (2.104)]}, \tag{2.110}$$

$$E_y = \frac{1}{j\omega\varepsilon_0\varepsilon_r} \left(-j\beta H_x - \frac{\partial H_z}{\partial x}\right) \quad \text{[from Eq. (2.105)]}, \tag{2.11}$$

$$E_z = \frac{-1}{j\omega\varepsilon_0\varepsilon_r} \frac{\partial H_x}{\partial y} \quad \text{[from Eq. (2.106)]}. \tag{2.112}$$

On the other hand, the component representation of the magnetic divergence equation

$$\nabla \cdot \mathbf{H} = 0 \tag{2.113}$$

is expressed as

$$\frac{\partial H_x}{\partial x} - j\beta H_z = 0. \tag{2.114}$$

32 ANALYTICAL METHODS

Thus, we get the longitudinal magnetic field component

$$H_z = \frac{1}{j\beta} \frac{\partial H_x}{\partial x}. \tag{2.115}$$

Eliminating H_z by substituting Eq. (2.115) into Eqs. (2.110) and (2.111), we get

$$E_x = \frac{1}{j\omega\varepsilon_0\varepsilon_r} \frac{\partial}{\partial y}\left(\frac{1}{j\beta}\frac{\partial H_x}{\partial x}\right) = \frac{-1}{\omega\varepsilon_0\varepsilon_r\beta} \frac{\partial^2 H_x}{\partial x \partial y} \quad \text{[from Eq. (2.110)]}, \tag{2.116}$$

$$E_y = \frac{1}{j\omega\varepsilon_0\varepsilon_r}\left[-j\beta H_x - \frac{\partial}{\partial x}\left(\frac{1}{j\beta}\frac{\partial H_x}{\partial x}\right)\right]$$

$$= -\frac{1}{\omega\varepsilon_0\varepsilon_r\beta}\left(\beta^2 H_x - \frac{\partial^2 H_x}{\partial x^2}\right) \quad \text{[from Eq. (2.111)]}. \tag{2.117}$$

The above results can be summarized as

$$E_x = \frac{-1}{\omega\varepsilon_0\varepsilon_r\beta} \frac{\partial^2 H_x}{\partial x \partial y} \quad \text{[from Eq. (2.116)]}, \tag{2.118}$$

$$E_y = -\frac{1}{\omega\varepsilon_0\varepsilon_r\beta}\left(\beta^2 H_x - \frac{\partial^2 H_x}{\partial x^2}\right) \quad \text{[from Eq. (2.117)]}, \tag{2.119}$$

$$E_z = \frac{-1}{j\omega\varepsilon_0\varepsilon_r} \frac{\partial H_x}{\partial y} \quad \text{[from Eq. (2.112)]}, \tag{2.120}$$

$$H_z = \frac{1}{j\beta} \frac{\partial H_x}{\partial x} \quad \text{[from Eq. (2.115)]}. \tag{2.121}$$

Substituting Eqs. (2.119) and (2.120) into Eq. (2.107), we get the following wave equation for a principal field component H_x for the E_{pq}^y mode:

$$\frac{\partial^2 H_x}{\partial x^2} + \frac{\partial^2 H_x}{\partial y^2} + k_0^2(\varepsilon_r - n_{\text{eff}}^2)H_x = 0. \tag{2.122}$$

The electric field and the magnetic field components to be connected are H_x and E_z at the boundaries $y = \pm b$ and are H_z and E_y at the boundaries $x = \pm a$.

Since the principal field component H_x is a solution (i.e., a wave function) of the wave equation (2.122), we get

$$H_x = C_1 \cos(k_x x + \alpha_1) \cos(k_y y + \alpha_2) \quad \text{(region 1)}, \tag{2.123}$$

$$= C_2 \cos(k_x x + \alpha_1) \exp[-\gamma_y(y - b)] \quad \text{(region 2)}, \tag{2.124}$$

$$= C_3 \exp[-\gamma_x(x - a)] \cos(k_y y + \alpha_2) \quad \text{(region 3)}, \tag{2.125}$$

$$= C_4 \cos(k_x x + \alpha_1) \exp[\gamma_y(y + b)] \quad \text{(region 4)}, \tag{2.126}$$

$$= C_5 \exp[\gamma_x(x + a)] \cos(k_y y + \alpha_2) \quad \text{(region 5)}, \tag{2.127}$$

Substituting these wave functions into the wave equation (2.122), we get (after some mathematical manipulations) for the wave numbers the same equations specified by Eqs. (2.81)–(2.85) for the E^x_{pq} mode. Equations (2.96) and (2.97) also hold.

The next step is to impose the boundary conditions on the tangential electric fields and the magnetic field components at the interfaces between different media.

A. Connection at $y = \pm b$: E_z and H_x Setting the electric field components E_z of regions 1 and 2, obtained by substituting Eqs. (2.123) and (2.124) into Eq. (2.120), equal at $y = b$, we get

$$C_1 \frac{k_y}{\varepsilon_{r1}} \sin(k_y b + \alpha_2) = C_2 \frac{\gamma_y}{\varepsilon_{r2}}. \tag{2.128}$$

And setting the electric fields H_x of regions 1 and 2 equal at $y = b$, from Eqs. (2.123) and (2.124), we get

$$C_1 \cos(k_y b + \alpha_2) = C_2. \tag{2.129}$$

Dividing Eq. (2.128) by Eq. (2.129), we get

$$\tan(k_y b + \alpha_2) = \frac{\varepsilon_{r1} \gamma_y}{\varepsilon_{r2} k_y}.$$

34 ANALYTICAL METHODS

Therefore

$$k_y b + \alpha_2 = \tan^{-1}\left(\frac{\varepsilon_{r1}\gamma_y}{\varepsilon_{r2}k_y}\right) + q_1\pi \quad (q_1 = 0, 1, \ldots). \tag{2.130}$$

On the other hand, setting the electric field components E_z of regions 1 and 4, which are obtained by substituting Eqs. (2.123) and (2.126) into Eq. (2.120), equal at $y = -b$, we get

$$C_1 \frac{k_y}{\varepsilon_{r1}} \sin(-k_y b + \alpha_2) = -C_4 \frac{\gamma_y}{\varepsilon_{r2}}. \tag{2.131}$$

Setting the magnetic field components H_x of regions 1 and 4 equal at $y = -b$, from Eqs. (2.123) and (2.126), we get

$$C_1 \cos(-k_y b + \alpha_2) = C_4. \tag{2.132}$$

Dividing Eq. (2.131) by Eq. (2.132), we get

$$\tan(k_y b - \alpha_2) = \frac{\varepsilon_{r1}\gamma_y}{\varepsilon_{r2}k_y}.$$

Therefore

$$k_y b - \alpha_2 = \tan^{-1}\left(\frac{\varepsilon_{r1}\gamma_y}{\varepsilon_{r2}k_y}\right) + q_2\pi \quad (q_2 = 0, 1, \ldots). \tag{2.133}$$

Adding Eq. (2.130) to Eq. (2.133), we get

$$k_y b = \tan^{-1}\left(\frac{\varepsilon_{r1}\gamma_y}{\varepsilon_{r2}k_y}\right) + \tfrac{1}{2}(q-1)\pi \quad (q = 1, 2, \ldots) \tag{2.134}$$

B. Connection at $x = \pm a$: E_y and H_z Setting the electric field components H_z of regions 1 and 3, obtained by substituting Eqs. (2.123) and (2.125) into Eq. (2.119), equal at $x = a$, we get

$$C_1 \frac{\beta^2 + k_x^2}{\varepsilon_{r1}} \cos(k_x a + \alpha_1) = \frac{\beta^2 - \gamma_x^2}{\varepsilon_{r2}} C_3. \tag{2.135}$$

2.3 MARCATILI'S METHOD

Setting the electric field components H_z of regions 1 and 3, obtained by substituting Eqs. (2.123) and (2.125) into Eq. (2.121), equal at $x = a$, we get

$$C_1 k_x \sin(k_x a + \alpha_1) = C_3 \gamma_x. \tag{2.136}$$

Dividing Eq. (2.136) by Eq. (2.135), we get

$$\tan(k_x a + \alpha_1) = \frac{\varepsilon_{r2}(\beta^2 + k_x^2)\gamma_x}{\varepsilon_{r1}(\beta^2 - \gamma_x^2)k_x}. \tag{2.137}$$

Here, making use of Eqs. (2.96) and (2.97), we get

$$\tan(k_x a + \alpha_1) = \frac{\varepsilon_{r2}(k_0^2 n_1^2 - k_y^2)\gamma_x}{\varepsilon_{r1}(k_0^2 n_2^2 - k_y^2)k_x}.$$

Therefore

$$k_x a + \alpha_1 = \tan^{-1}\left(\frac{\varepsilon_{r2}(k_0^2 n_1^2 - k_y^2)\gamma_x}{\varepsilon_{r1}(k_0^2 n_2^2 - k_y^2)k_x}\right) + p_1 \pi \quad (p_1 = 0, 1, \ldots). \tag{2.138}$$

On the other hand, setting the electric field components E_y of regions 1 and 5, obtained by substituting Eqs. (2.123) and (2.127) into Eq. (2.119), equal at $x = -a$, we get

$$C_1 \frac{\beta^2 + k_x^2}{\varepsilon_{r1}} \cos(-k_x a + \alpha_1) = \frac{\beta^2 - \gamma_x^2}{\varepsilon_{r2}} C_5. \tag{2.139}$$

And setting the magnetic field components H_z of regions 1 and 5, obtained by substituting Eqs. (2.123) and (2.127) into Eq. (2.121), equal at $x = -a$, we get

$$C_1 k_x \sin(k_x a - \alpha_1) = C_3 \gamma_x. \tag{2.140}$$

Dividing Eq. (2.140) by Eq. (2.139), we get

$$\tan(k_x a - \alpha_1) = \frac{\varepsilon_{r2}(\beta^2 + k_x^2)\gamma_x}{\varepsilon_{r1}(\beta^2 - \gamma_x^2)k_x}.$$

And substituting Eqs. (2.96) and (2.97) into this equation, we get

$$\tan(k_x a - \alpha_1) = \frac{\varepsilon_{r2}(k_0^2 n_1^2 - k_y^2)\gamma_x}{\varepsilon_{r1}(k_0^2 n_2^2 - k_y^2)k_x}.$$

Therefore

$$k_x a - \alpha_1 = \tan^{-1}\left(\frac{\varepsilon_{r2}(k_0^2 n_1^2 - k_y^2)\gamma_x}{\varepsilon_{r1}(k_0^2 n_2^2 - k_y^2)k_x}\right) + p_2\pi \qquad (p_2 = 0, 1, \ldots). \tag{2.141}$$

Here, adding Eq. (2.138) to Eq. (2.141), we get

$$k_x a = \tan^{-1}\left(\frac{\varepsilon_{r2}(k_0^2 n_1^2 - k_y^2)\gamma_x}{\varepsilon_{r1}(k_0^2 n_2^2 - k_y^2)k_x}\right) + \tfrac{1}{2}(p - 1)\pi \qquad (p = 1, 2, \ldots). \tag{2.142}$$

Or, since $k_0 n_{1,2} \gg k_y$ for most cases, we get

$$k_x a = \tan^{-1}\left(\frac{\gamma_x}{k_x}\right) + \tfrac{1}{2}(p - 1)\pi \qquad (p = 1, 2, \ldots). \tag{2.143}$$

The propagation constant β (or the effective index n_{eff}) is obtained in just the same way as for the E^x_{pq} mode.

Equation (2.134) corresponds to the characteristic equation of the TM mode for a three-layer slab optical waveguide parallel to the x axis. Equation (2.143), on the other hand, corresponds to the characteristic equation of the TE mode for a three-layer slab optical waveguide parallel to the y axis. The correspondence is the same as that described in the last part of Section 2.2.

2.4 METHOD FOR AN OPTICAL FIBER

In this section, we discuss an analysis of a step-index optical fiber. An optical fiber consists of a core and a cladding and is axially symmetric. Since the refractive index of the core is slightly higher than that of the cladding, the optical field is largely confined to the core. A single-mode

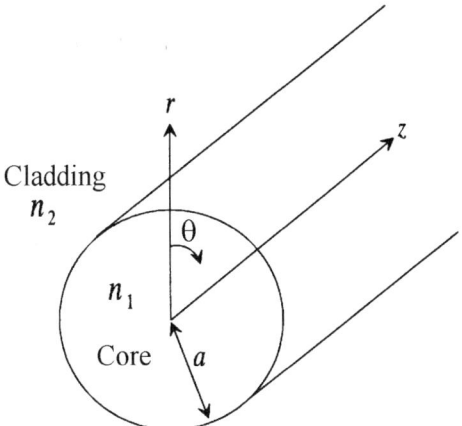

FIGURE 2.4. Step-index optical fiber.

fiber, which has only one guided mode, plays a key role in telecommunications systems.

Figure 2.4 shows a cross-sectional view of a step-index optical fiber. The core of a radius a has a uniform refractive index n_1, slightly higher than the refractive index of the cladding, n_2. Thus, the relative permittivities of the core and cladding are respectively $\varepsilon_{r1} = n_1^2$ and $\varepsilon_{r2} = n_2^2$.

The exact vectorial wave equations for the electric field and the magnetic field were shown in Chapter 1 as Eqs. (1.30) and (1.36). Since the structure of the optical fiber is uniform in the propagation direction, we can, as shown in Eq. (1.39), substitute $-j\beta$ for the derivatives of the electric and magnetic fields with respect to z.

Thus, we can write

$$\nabla_\perp^2 \mathbf{E} + \nabla\left(\frac{\nabla \varepsilon_r}{\varepsilon_r} \cdot \mathbf{E}\right) + k_0^2(\varepsilon_r - n_{\text{eff}}^2)\mathbf{E} = 0, \qquad (2.144)$$

$$\nabla_\perp^2 \mathbf{H} \frac{\nabla \varepsilon_r}{\varepsilon_r} \times (\nabla \times \mathbf{H}) + k_0^2(\varepsilon_r - n_{\text{eff}}^2)\mathbf{H} = 0, \qquad (2.145)$$

where $\mathbf{E}$ and $\mathbf{H}$ are respectively the electric fields and the magnetic fields and n_{eff} is the effective index to be obtained.

Although the structure shown in Fig. 2.4 can be analyzed exactly by using the hybrid-mode analysis [4], the analysis procedure is a little complicated. Fortunately, however, the weakly guiding approximation can be used because the refractive index difference between the core and the cladding is very small, about 1%. This approximation simplifies the analysis significantly, and the modes obtained are called linearly polarized modes [5].

2.4.1 Linearly Polarized Modes (LP Modes)

A. Field Expressions Since the derivative of the relative permittivity ε_r is small, neglecting the derivatives in the vectorial wave equations (2.144) and (2.145) gives a good approximation. Using this approximation, we can reduce the vectorial wave equations to the scalar Helmholtz equations for the tangential electric fields and the magnetic fields:

$$\nabla_\perp^2 \mathbf{E}_\perp + k_0^2(\varepsilon_r - n_{\text{eff}}^2)\mathbf{E}_\perp = \mathbf{0}, \qquad (2.146)$$

$$\nabla_\perp^2 \mathbf{H}_\perp + k_0^2(\varepsilon_r - n_{\text{eff}}^2)\mathbf{H}_\perp = \mathbf{0}. \qquad (2.147)$$

Since the optical fiber is axially symmetric, the Laplacian ∇^2 is rewritten for a cylindrical coordinate system as follows:

$$\begin{aligned}\nabla^2 &= \nabla_\perp^2 + \frac{\partial^2}{\partial z^2} \\ &= \frac{1}{r}\frac{\partial}{\partial r}\left(r\frac{\partial}{\partial r}\right) + \frac{1}{r^2}\frac{\partial^2}{\partial \theta^2} + \frac{\partial^2}{\partial z^2} \\ &= \frac{\partial^2}{\partial r^2} + \frac{1}{r}\frac{\partial}{\partial r} + \frac{1}{r^2}\frac{\partial^2}{\partial \theta^2} + \frac{\partial^2}{\partial z^2}.\end{aligned}$$

To solve Eq. (2.146), we assume that the tangential electric field component $\mathbf{E}_\perp$ (i.e., E_x or E_y) is given by

$$E_i(r, \theta) = R(r)\Theta(\theta), \qquad (2.148)$$

where the subscript $i = x, y$.

Substituting the electric field component (2.148) into Eq. (2.146) and dividing the resultant equation by $R(r)\Theta(\theta)$, we get

$$\frac{1}{R(r)}\left(\frac{\partial^2 R(r)}{\partial r^2} + \frac{1}{r}\frac{\partial R(r)}{\partial r}\right) + \frac{1}{r^2}\frac{1}{\Theta(\theta)}\frac{\partial^2 \Theta(\theta)}{\partial \theta^2} + k_0^2(\varepsilon_r - n_{\text{eff}}^2) = 0.$$

Therefore

$$\frac{r^2}{R(r)}\left(\frac{\partial^2 R(r)}{\partial r^2}+\frac{1}{r}\frac{\partial R(r)}{\partial r}\right)+r^2 k_0^2(\varepsilon_r - n_{\text{eff}}^2) = -\frac{1}{\Theta(\theta)}\frac{\partial^2 \Theta(\theta)}{\partial \theta^2}. \quad (2.149)$$

Since the left-hand side of Eq. (2.149) is a function of only the variable r and the right-hand side is a function of only the variable θ, both sides have to be constant. Thus, we get

$$\frac{r^2}{R(r)}\left(\frac{d^2 R(r)}{dr^2}+\frac{1}{r}\frac{dR(r)}{dr}\right)+r^2 k_0^2(\varepsilon_r - n_{\text{eff}}^2) = l^2 \quad (2.150)$$

and

$$\frac{1}{\Theta(\theta)}\frac{d^2 \Theta(\theta)}{d\theta^2} = -l^2. \quad (2.151)$$

Equations (2.149) and (2.151) are summarized as

$$\frac{d^2 R(r)}{dr^2}+\frac{1}{r}\frac{dR(r)}{dr}+k_0^2\left(\varepsilon_r - n_{\text{eff}}^2 - \frac{l^2}{r^2}\right)R(r) = 0 \quad (2.152)$$

and

$$\frac{d^2 \Theta(\theta)}{d\theta^2}+l^2\Theta(\theta) = 0. \quad (2.153)$$

The solution of Eq. (2.153) is an oscillation with a single frequency and is expressed as

$$\Theta(\theta) = \sin(l\theta + \phi), \quad (2.154)$$

where l and ϕ are respectively an integer and an arbitrary constant phase. The next step is to solve Eq. (2.152). Using the variable transformations

$$\tilde{u}^2 = k_0^2(\varepsilon_r - n_{\text{eff}}^2) \quad (2.155)$$

and

$$\xi = \tilde{u}r, \quad (2.156)$$

we write the first and second derivatives with respect to r as

$$\frac{d}{dr} = \frac{d}{d\xi}\frac{d\xi}{dr} = \tilde{u}\frac{d}{d\xi} \quad \text{and} \quad \frac{d^2}{dr^2} = \tilde{u}\frac{d^2}{d\xi^2}\frac{d\xi}{dr} = \tilde{u}^2\frac{d^2}{d\xi^2}. \tag{2.157}$$

Equation (2.152) is then rewritten as

$$\tilde{u}^2\frac{d^2R(r)}{d\xi^2} + \frac{1}{\xi/\tilde{u}}\tilde{u}\frac{dR(r)}{d\tilde{u}} + k_0^2\left(\tilde{u}^2 - \frac{l^2}{(\xi/\tilde{u})^2}\right)R(r) = 0.$$

Therefore

$$\frac{d^2R(r)}{d\xi^2} + \frac{1}{\xi}\frac{dR(r)}{d\xi} + k_0^2\left(1 - \frac{l^2}{\xi^2}\right)R(r) = 0. \tag{2.158}$$

Solutions for Eq. (2.158) are lth-order Bessel functions and are written as

$$R(r) = \begin{cases} AJ_l\left(\frac{ur}{a}\right) + BN_l\left(\frac{ur}{a}\right) & \text{for } r \leq a, \\ CK_l\left(\frac{wr}{a}\right) + DI_l\left(\frac{wr}{a}\right) & \text{for } r \geq a. \end{cases} \tag{2.159}$$

Here, $J_l(ur/a)$ and $N_l(ur/a)$ are the lth-order Bessel functions of the first and second kinds, and $K_l(wr/a)$ and $I_l(wr/a)$ are the lth-order modified Bessel functions of the first and second kinds. The parameters u^2 and w^2 are defined as

$$u^2 = k_0^2 a^2(\varepsilon_{r1} - n_{\text{eff}}^2), \tag{2.160}$$
$$w^2 = k_0^2 a^2(n_{\text{eff}}^2 - \varepsilon_{r2}). \tag{2.161}$$

Thus, we get the following very important relation between u and w:

$$u^2 + w^2 = v^2, \tag{2.162}$$

where

$$v = k_0 a\sqrt{\varepsilon_{r1} - \varepsilon_{r2}} \tag{2.163}$$

is the normalized frequency. The parameters u and w are respectively considered to be the normalized lateral propagation constant in the core and the normalized lateral decay constant in the cladding. The effective index n_{eff} has to satisfy the relation

$$n_1 \geq n_{\text{eff}} \geq n_2. \tag{2.164}$$

Since the Bessel function of the second kind, $N_l(ur/a)$, diverges at $r = 0$ and the modified Bessel function of the second kind, $I_l(wr/a)$, diverges at $r = \infty$, the coefficients B and D of those functions have to be zero. Thus, from Eq. (2.159), we get

$$R(r) = \begin{cases} AJ_l\left(\dfrac{ur}{a}\right) & \text{for } r \leq a, \\ CK_l\left(\dfrac{wr}{a}\right) & \text{for } r \geq a. \end{cases} \tag{2.165}$$

B. Characteristic Equation The boundary conditions to be satisfied by the radial wave function $R(r)$ are

$$R(a - 0) = R(a + 0) \tag{2.166}$$

and

$$\left.\frac{dR(r)}{dr}\right|_{a-0} = \left.\frac{dR(r)}{dr}\right|_{a+0}. \tag{2.167}$$

From Eqs. (2.166) and (2.167), we get

$$AJ_l(u) - CK_l(w) = 0, \tag{2.168}$$

$$AuJ_l'(u) - CwK_l'(w) = 0. \tag{2.169}$$

Equations (2.166) and (2.167) can be rewritten as the matrix equation

$$\begin{pmatrix} J_l(u) & -K_l(w) \\ uJ_l'(u) & -wK_l'(w) \end{pmatrix} \begin{pmatrix} A \\ C \end{pmatrix} = 0. \tag{2.170}$$

When the coefficients A and C are nontrivial, the determinant of their coefficient matrix has to be zero such that

$$\begin{vmatrix} J_l(u) & -K_l(w) \\ uJ_l'(u) & -wK_l'(w) \end{vmatrix} = 0. \qquad (2.171)$$

Equation (2.171) can be rewritten as

$$-wJ_l(u)K_l'(w) + uJ_l'(u)K_l(w) = 0,$$

where the prime denotes the derivative with respect to r.

Thus, we get the well-known characteristic equation

$$\frac{uJ_l'(u)}{J_l(u)} = \frac{wK_l'(w)}{K_l(w)}. \qquad (2.172)$$

The effective index n_{eff} can be obtained by solving a combination of Eqs. (2.162) and (2.172). In other words, since u and w are functions of the effective index n_{eff}, the characteristic equation (2.172) is a function of n_{eff}. Solutions of this equation are called linearly polarized modes (LP modes).

1. Explicit Forms of the Characteristic Equation

LP_{0m} MODES ($l = 0$ AND $m \geq 1$) Equation (2.172) is rewritten for $l = 0$ as

$$\frac{uJ_0'(u)}{J_0(u)} = \frac{wK_0'(w)}{K_0(w)}. \qquad (2.173)$$

Making use of the Bessel function formulas

$$J_0'(z) = -J_1(z), \qquad (2.174)$$

$$K_0'(z) = -K_1(z), \qquad (2.175)$$

we rewrite Eq. (2.173) as

$$\frac{uJ_1(u)}{J_0(u)} = \frac{wK_1(w)}{K_0(w)}.$$

Therefore

$$\frac{J_0(u)}{uJ_1(u)} = \frac{K_0(w)}{wK_1(w)}. \tag{2.176}$$

LP$_{lm}$ MODES ($l \geq 1$ AND $m \geq 1$) Here, we consider the LP$_{lm}$ mode. Equation (2.172) is rewritten for $l = 1$ as

$$\frac{uJ_1'(u)}{J_1(u)} = \frac{wK_1'(w)}{K_1(w)}. \tag{2.177}$$

Making use of the Bessel function formulas

$$J_1'(z) = J_0(z) - z^{-1}J_1(z), \tag{2.178}$$

$$zK_1'(z) = -zK_0(z) - K_1(z), \tag{2.179}$$

we rewrite Eq. (2.177) as

$$\frac{u(J_0(u) - u^{-1}J_1(u))}{J_1(u)} = \frac{-wK_0(w) - K_1(w)}{K_1(w)}.$$

Therefore

$$\frac{J_1(u)}{uJ_0(u)} = -\frac{K_1(w)}{wK_0(w)}. \tag{2.180}$$

In general, making use of the Bessel function formulas

$$J_\nu'(z) = J_{\nu-1}(z) - \nu z^{-1}J_\nu(z), \tag{2.181}$$

$$zK_\nu'(z) = -zK_{\nu-1}(z) - \nu K_\nu(z), \tag{2.182}$$

one can rewrite Eq. (2.177) as

$$\frac{u(J_{l-1}(u) - lu^{-1}J_l(u))}{J_l(u)} = \frac{-wK_{l-1}(w) - lK_l(w)}{K_l(w)}.$$

Therefore

$$\frac{J_l(u)}{uJ_{l-1}(u)} = -\frac{K_l(w)}{wK_{l-1}(w)}. \quad (2.183)$$

2. Value Range of u The characteristic equations for LP$_{lm}$ modes have solutions only within limited ranges of the parameter u, and we need to know what these ranges are. These ranges can be determined by investigating the limits $w \to 0$ and $w \to +\infty$, where the former corresponds to $u \to v$. Through this process, the single-mode condition and cutoff conditions for the higher order LP$_{lm}$ modes will also be clarified.

LP$_{0m}$ MODES ($l = 0$ AND $m \geq 1$) First, we investigate the limit $w \to 0$ (i.e., $u \to v$). Since the zeroth-order and lth-order modified Bessel functions of the first kind can, for the limit of $z \to 0$, be respectively expressed asymptotically as

$$K_0(z) \sim -\ln z, \quad (2.184)$$

$$K_l(z) \sim \tfrac{1}{2}\Gamma(l)(\tfrac{1}{2}z)^{-1} \quad \text{for } l > 0. \quad (2.185)$$

The right-hand side of Eq. (2.176) can be rewritten as

$$\frac{K_0(w)}{wK_1(w)} = \frac{\ln w}{w(1/2)\Gamma(1)[(1/2)w]^{-1}} = -\ln w \to +\infty \quad \text{for } w \to 0.$$

The left-hand side of Eq. (2.176) also has to go to $+\infty$. That is,

$$\frac{J_0(v)}{vJ_1(v)} \to +\infty. \quad (2.186)$$

The possible solutions for Eq. (2.186) are $v \to 0$ and $J_1(v) \to 0$. Since we get $J_0(v) \to 1$ and $J_1(v) \to +0$ for $v \to 0$, Eq. (2.186) holds. This implies that the cutoff value v_c for the normalized frequency v is zero. In other words, the LP$_{01}$ mode has no cutoff condition. Next, we discuss the cutoff conditions for higher order modes: LP$_{0m}$ modes where $m \geq 2$. Here, we assume that $j_{1,m-1}$ is the $(m-1)$th zero of the Bessel function of the first kind. That is, $J_1(j_{1,m-1}) = 0$. Since the signs of $J_0(v)$ and $J_1(v)$ are the same for the limit $v \to j_{1,m-1} + 0$, Eq. (2.186) holds. This means the cutoff value v_c of the LP$_{0m}$ mode is given by

$$v_c = j_{1,m-1}. \tag{2.187}$$

Thus, we can summarize the cutoff conditions for LP$_{0m}$ modes as follows:

$$\begin{aligned}\text{LP}_{01} \text{ mode:} &\quad v_c = 0, \\ \text{LP}_{0m} \text{ mode:} &\quad v_c = j_{1,m-1} \quad \text{for } m \geq 2.\end{aligned} \tag{2.188}$$

On the other hand, when $w \gg 1$, the asymptotic expansion of the lth-order modified Bessel function of the first kind is

$$\begin{aligned} K_l &\sim \sqrt{\frac{\pi}{2}} e^{-w} \sum_{n=0}^{\infty} \frac{(l,n)}{(2w)^n} \\ &\sim \sqrt{\frac{\pi}{2}} e^{-w} \frac{(l,0)}{(2w)^0} \\ &\sim \sqrt{\frac{\pi}{2}} e^{-w}. \end{aligned} \tag{2.189}$$

Since for $w \gg 1$ the right-hand side of Eq. (2.176) can be rewritten as

$$\frac{K_0(w)}{wK_1(w)} \sim \frac{\sqrt{\pi/2}\, e^{-w}}{w\sqrt{\pi/2}\, e^{-w}} = \frac{1}{w} \to 0, \tag{2.190}$$

the left-hand side of Eq. (2.176) also has to go to zero. That is,

$$\frac{J_0(u)}{uJ_1(u)} \to 0. \tag{2.191}$$

This implies that the asymptotic value of u is given by

$$u \sim j_{0,m}. \tag{2.192}$$

Thus, we can summarize the asymptotic values of u for LP_{0m} modes as follows:

$$\begin{aligned} LP_{01} \text{ mode:} \quad & u \sim j_{0,1}, \\ LP_{0m} \text{ mode:} \quad & u \sim j_{0,m} \quad \text{for } m \geq 2. \end{aligned} \quad (2.193)$$

LP_{lm} MODES ($l \geq 1$ AND $m \geq 1$) As in the above discussion, we first investigate the limit $w \to 0$ (i.e., $u \to v$). For the limit $z \to 0$, the lth-order modified Bessel functions of the first kind were shown in Eqs. (2.185) and (2.185). The $(l-1)$th-order modified Bessel function of the first kind for the limit $z \to 0$ is

$$K_{l-1}(z) \sim \tfrac{1}{2}\Gamma(l-1)(\tfrac{1}{2}z)^{-l+1} \quad \text{for } z \to 0. \quad (2.194)$$

Making use of this approximation, we express the right-hand side of Eq. (2.183) as

$$\begin{aligned} -\frac{K_l(w)}{wK_{l-1}(w)} &= -\frac{1}{w}\frac{(1/2)\Gamma(l)[(1/2)w]^{-l}}{(1/2)\Gamma(l-1)[(1/2)w]^{-l+1}} \\ &= -\frac{1}{w}\frac{(l-1)![(1/2)w]^{-l}}{(l-2)![(1/2)w]^{-l+1}} \\ &= -\frac{1}{w}(l-1)\frac{2}{w} \\ &= -\frac{2(l-1)}{w^2} \to -\infty \quad \text{for } w \to 0. \quad (2.195) \end{aligned}$$

The left-hand side of Eq. (2.183) also has go to $-\infty$. That is,

$$\frac{J_l(v)}{vJ_{l-1}(v)} \to -\infty. \quad (2.196)$$

The possible solutions of Eq. (2.196) are $v \to 0$ and $J_{l-1}(v) \to 0$. Since the left-hand side of Eq. (2.183) diverges to $+\infty$ for the limit $v \to 0$, the limit $v \to 0$ cannot be a solution. On the other hand, since the signs of $J_l(v)$ and $J_{l-1}(v)$ are opposite for $v \to j_{l-1,m} + 0$, $j_{l-1,m}$ can be a solution for Eq. (2.196). Here, $j_{l-1,m}$ is the zero of the $(l-1)$th-order Bessel function of the first kind, $J_{l-1}(v)$.

The cutoff condition for LP$_{lm}$ modes is

$$\text{LP}_{lm} \text{ mode:} \quad v_c = j_{l-1,m}. \quad (2.197)$$

On the other hand, since, according to the asymptotic expression (2.189), the right-hand side of Eq. (2.183) is 0 for $w \gg 1$, the left-hand side of Eq. (2.176) also has to go to zero. That is,

$$\frac{J_l(u)}{uJ_{l-1}(u)} \to 0. \quad (2.198)$$

This implies that the asymptotic value of u is given by

$$u \approx j_{l,m}. \quad (2.199)$$

We can thus summarize the possible value ranges of the parameter u as

$$\begin{aligned}
&\text{LP}_{01} \text{ mode:} & & 0 \leq u < j_{0,1}, \\
&\text{LP}_{0m} \text{ mode:} & & j_{1,m-1} \leq u < j_{0,m} \quad \text{for } m \geq 2, \\
&\text{LP}_{lm} \text{ mode:} & & j_{l-1,m} \leq u < j_{l,m} \quad \text{for } l \geq 1, \ m \geq 1.
\end{aligned} \quad (2.200)$$

It should be noted that since the minimum value of the zeros of the Bessel functions is $j_{0,1}$ (2.404826), the single-mode condition for an optical fiber with a step index is given by the cutoff value for the LP$_{11}$ mode:

$$v_c = j_{0,1}. \quad (2.201)$$

Figure 2.5 shows examples of u–ω curves. The crossing points of the curve $u^2 + w^2 = v^2$ and the other curves give the effective indexes and confirm the value ranges of u specified above.

Figure 2.6 shows examples of calculated field distributions. It should be noted that l and m of LP$_{lm}$ respectively correspond to the number of dark lines in the azimuthal direction and the number of bright peaks in the radial direction.

2.4.2 Hybrid-Mode Analysis

This section discusses a more exact analysis for the step-index optical fiber. Since the above LP-mode analysis can meet the ordinary demands for fiber analyses and the hybrid-mode analyses are very specialized, some readers may wish to skip this section.

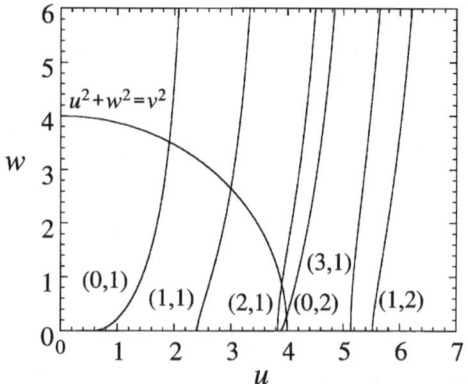

FIGURE 2.5. Relation between u and w ($v = 4$).

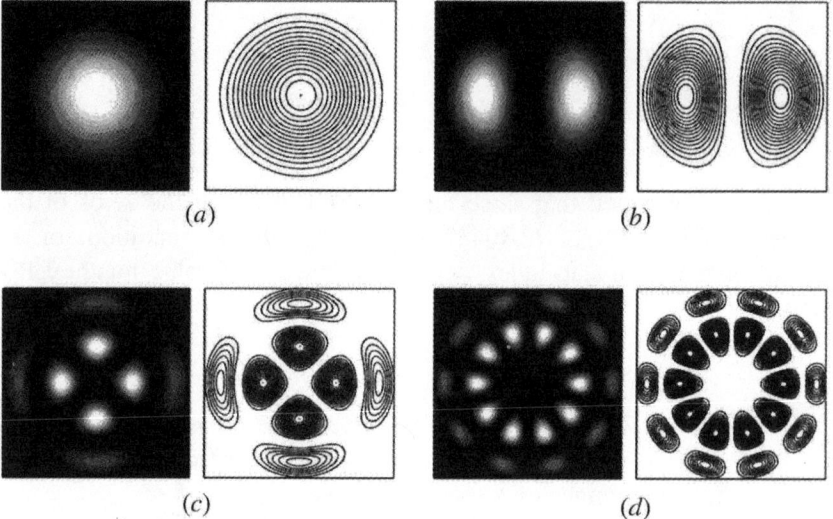

FIGURE 2.6. Field distributions for the LP mode ($v = 20$): (a) $LP_{0,1}$; (b) $LP_{1,1}$; (c) $LP_{2,2}$; (d) $LP_{5,2}$.

A. Field Expressions

The cylindrical electric field **E** and the cylindrical magnetic field **H** are expressed as

$$\mathbf{E}(r, \theta, z) = \mathbf{E}(r, \theta) \exp[j(\omega t - \beta t)], \quad (2.202)$$

$$\mathbf{H}(r, \theta, z) = \mathbf{H}(r, \theta) \exp[j(\omega t - \beta t)], \quad (2.203)$$

where

$$\mathbf{E}_0 = E_r\mathbf{r} + E_\theta\boldsymbol{\theta} + E_z\mathbf{z}, \tag{2.204}$$

$$\mathbf{H}_0 = H_r\mathbf{r} + H_\theta\boldsymbol{\theta} + H_z\mathbf{z}. \tag{2.205}$$

Here, $\mathbf{r}$, $\boldsymbol{\theta}$, and $\mathbf{z}$ are respectively unit vectors in the radial, azimuthal, and longitudinal directions. Applying the rotation formula

$$\nabla \times \mathbf{A} = \left(\frac{1}{r}\frac{\partial A_z}{\partial \theta} - \frac{\partial A_\theta}{\partial z}\right)\mathbf{r} + \left(\frac{\partial A_r}{\partial z} - \frac{\partial A_z}{\partial r}\right)\boldsymbol{\theta} + \left(\frac{1}{r}\frac{\partial}{\partial r}(rA_\theta) - \frac{1}{r}\frac{\partial A_r}{\partial \theta}\right)\mathbf{z} \tag{2.206}$$

for a vector $\mathbf{A} = A_r\mathbf{r} + A_\theta\boldsymbol{\theta} + A_z\mathbf{z}$ to the Maxwell equations

$$\nabla \times \mathbf{E} = -j\omega\mu_0\mathbf{H}, \tag{2.207}$$

$$\nabla \times \mathbf{H} = -j\omega\varepsilon_0\varepsilon_r\mathbf{E} = -j\omega\varepsilon\mathbf{E}, \tag{2.208}$$

we get

$$\frac{1}{r}\frac{\partial E_z}{\partial \theta} + j\beta E_\theta = -j\omega\mu_0 H_r, \tag{2.209}$$

$$-j\beta E_r - \frac{\partial E_z}{\partial r} = -j\omega\mu_0 H_\theta, \tag{2.210}$$

$$\frac{1}{r}\frac{\partial}{\partial r}(rE_\theta) - \frac{1}{r}\frac{\partial E_r}{\partial \theta} = -j\omega\mu_0 H_z, \tag{2.211}$$

$$\frac{1}{r}\frac{\partial H_z}{\partial \theta} + j\beta H_\theta = j\omega\varepsilon E_r, \tag{2.212}$$

$$-j\beta H_r - \frac{\partial H_z}{\partial r} = -j\omega\varepsilon E_\theta, \tag{2.213}$$

$$\frac{1}{r}\frac{\partial}{\partial r}(rH_\theta) - \frac{1}{r}\frac{\partial H_r}{\partial \theta} = -j\omega\varepsilon E_z. \tag{2.214}$$

Expressing the tangential field components (E_r, E_θ, H_r, and H_θ) as functions of the longitudinal field components (E_z and H_z), we get

$$j\beta E_\theta - j\omega\mu_0 H_r = \frac{1}{r}\frac{\partial E_z}{\partial \theta}, \qquad (2.215)$$

$$-j\beta E_r + j\omega\mu_0 H_\theta = \frac{\partial E_z}{\partial r}, \qquad (2.216)$$

$$j\omega\varepsilon E_r - j\beta H_\theta = \frac{1}{r}\frac{\partial H_z}{\partial \theta}, \qquad (2.217)$$

$$j\omega\varepsilon E_\theta + j\beta H_r = -\frac{\partial H_z}{\partial r}. \qquad (2.218)$$

The radial and azimuthal field components are obtained as follows:

Equation (2.216) × β + Eq. (2.217) × $\omega\mu_0$:

$$E_r = \frac{-j}{\omega^2\varepsilon\mu - \beta^2}\left(\beta\frac{\partial E_z}{\partial r} + \omega\mu_0\frac{1}{r}\frac{\partial H_z}{\partial \theta}\right). \qquad (2.219)$$

Equation (2.215) × β + Eq. (2.218) × $\omega\mu_0$:

$$E_\theta = \frac{-j}{\omega^2\varepsilon\mu - \beta^2}\left(\beta\frac{1}{r}\frac{\partial E_z}{\partial \theta} - \omega\mu_0\frac{\partial H_z}{\partial r}\right). \qquad (2.220)$$

Equation (2.215) × $\omega\varepsilon$ + Eq. (2.218) × β:

$$H_r = \frac{-j}{\omega^2\varepsilon\mu - \beta^2}\left(\beta\frac{\partial H_z}{\partial r} - \omega\varepsilon\frac{1}{r}\frac{\partial E_z}{\partial \theta}\right). \qquad (2.221)$$

Equation (2.216) × $\omega\varepsilon$ + Eq. (2.217) × β:

$$H_\theta = \frac{-j}{\omega^2\varepsilon\mu - \beta^2}\left(\beta\frac{1}{r}\frac{\partial H_z}{\partial \theta} + \omega\varepsilon\frac{\partial E_z}{\partial r}\right). \qquad (2.222)$$

Substituting the H_r of Eq. (2.221) and the H_θ of Eq. (2.222) into Eq. (2.214), we get

$$\frac{\partial^2 E_z}{\partial r^2} + \frac{1}{r}\frac{\partial E_z}{\partial r} + \frac{1}{r^2}\frac{\partial^2 E_z}{\partial \theta^2} + (\omega^2\varepsilon\mu - \beta^2)E_z = 0. \qquad (2.223)$$

2.4 METHOD FOR AN OPTICAL FIBER

And substituting the E_r of Eq. (2.219) and the E_θ of Eq. (2.220) into Eq. (2.211), we get

$$\frac{\partial^2 E_z}{\partial r^2} + \frac{1}{r}\frac{\partial E_z}{\partial r} + \frac{1}{r^2}\frac{\partial^2 E_z}{\partial \theta^2} + (\omega^2 \varepsilon \mu - \beta^2)E_z = 0. \qquad (2.224)$$

To solve Eqs. (2.223), we assume that the longitudinal field components E_z and H_z are given by

$$E_z(r, \theta)[\text{or } H_z(r, \theta)] = R_z(r)\Theta_z(\theta). \qquad (2.225)$$

Thus, we get the following wave equations for $R_z(r)$ and $\Theta_z(\theta)$:

$$\frac{d^2 R_z(r)}{dr^2} + \frac{1}{r}\frac{dR_z(r)}{dr} + k_0^2\left(\varepsilon_r - n_{\text{eff}}^2 - \frac{n^2}{r^2}\right)R_z(r) = 0 \qquad (2.226)$$

and

$$\frac{d^2\Theta_z(\theta)}{d\theta^2} + n^2\Theta_z(\theta) = 0. \qquad (2.227)$$

The solution of Eq. (2.227) is an oscillation with a single frequency and is expressed as

$$\Theta_z(\theta) = \sin(n\theta + \phi), \qquad (2.228)$$

where n and ϕ are respectively an integer and an arbitrary constant phase.

Through the procedure shown for the LP mode, we obtain the radial wave function $R_z(r)$ in the core as the Bessel function of the first kind, $J_n(ur/a)$, and obtain the radial wave function in the cladding as the modified Bessel function of the first kind, $K_n(ur/a)$. Thus, we finally get the following field components:

$$E_z = \begin{cases} AJ_n\left(\dfrac{ur}{a}\right)\sin(n\theta + \phi) & \text{for } r \leq a, \\ CK_n\left(\dfrac{wr}{a}\right)\sin(n\theta + \phi) & \text{for } r \geq a, \end{cases} \quad (2.229)$$

$$H_z = \begin{cases} BJ_n\left(\dfrac{ur}{a}\right)\cos(n\theta + \phi) & \text{for } r \leq a, \\ DK_n\left(\dfrac{wr}{a}\right)\cos(n\theta + \phi) & \text{for } r \geq a. \end{cases} \quad (2.230)$$

Substituting Eqs. (2.229) and (2.230) into Eqs. (2.219)–(2.222), we get the wave functions as follows:

1. In the core ($r \leq a$):

$$E_z = AJ_n\left(\dfrac{ur}{a}\right)\sin(n\theta + \phi), \quad (2.231)$$

$$E_r = \left[-A\dfrac{j\beta}{u/a}J_n'\left(\dfrac{ur}{a}\right) + B\dfrac{j\omega\mu_0}{(u/a)^2}\dfrac{n}{r}J_n\left(\dfrac{ur}{a}\right)\right]\sin(n\theta + \phi), \quad (2.232)$$

$$E_\theta = \left[-A\dfrac{j\beta}{(u/a)^2}\dfrac{n}{r}J_n\left(\dfrac{ur}{a}\right) + B\dfrac{j\omega\mu_0}{u/a}J_n'\left(\dfrac{ur}{a}\right)\right]\cos(n\theta + \phi),$$

$$(2.233)$$

$$H_z = BJ_n\left(\dfrac{ur}{a}\right)\cos(n\theta + \phi), \quad (2.234)$$

$$H_r = \left[A\dfrac{j\omega\varepsilon_1}{(u/a)^2}\dfrac{n}{r}J_n\left(\dfrac{ur}{a}\right) - B\dfrac{j\beta}{u/a}J_n'\left(\dfrac{ur}{a}\right)\right]\cos(n\theta + \phi), \quad (2.235)$$

$$H_\theta = \left[-A\dfrac{j\omega\varepsilon_1}{u/a}J_n'\left(\dfrac{ur}{a}\right) + B\dfrac{j\beta}{(u/a)^2}\dfrac{n}{r}J_n\left(\dfrac{ur}{a}\right)\right]\sin(n\theta + \phi).$$

$$(2.236)$$

2. In the cladding ($r \geq a$):

$$E_z = CK_n\left(\frac{wr}{a}\right)\sin(n\theta + \phi), \tag{2.237}$$

$$E_r = \left[C\frac{j\beta}{w/a}K_n'\left(\frac{wr}{a}\right) - D\frac{j\omega\mu_0}{(w/a)^2}\frac{n}{r}K_n\left(\frac{wr}{a}\right)\right]\sin(n\theta + \phi), \tag{2.238}$$

$$E_\theta = \left[C\frac{j\beta}{(w/a)^2}\frac{n}{r}K_n\left(\frac{wr}{a}\right) - D\frac{j\omega\mu_0}{w/a}K_n'\left(\frac{wr}{a}\right)\right]\cos(n\theta + \phi), \tag{2.239}$$

$$H_z = DK_n\left(\frac{wr}{a}\right)\cos(n\theta + \phi), \tag{2.240}$$

$$H_r = \left[-C\frac{j\omega\varepsilon_2}{(u/a)^2}\frac{n}{r}K_n\left(\frac{ur}{a}\right) + D\frac{j\beta}{w/a}K_n'\left(\frac{wr}{a}\right)\right]\cos(n\theta + \phi), \tag{2.241}$$

$$H_\theta = \left[C\frac{j\omega\varepsilon_2}{w/a}K_n'\left(\frac{wr}{a}\right) - D\frac{j\beta}{(w/a)^2}\frac{n}{r}K_n\left(\frac{wr}{a}\right)\right]\sin(n\theta + \phi). \tag{2.242}$$

The normalized lateral propagation constant u in the core and the normalized lateral decay constant w in the cladding were respectively defined in Eqs. (2.160) and (2.161).

B. Characteristic Equation The boundary conditions to be satisfied are that each of the tangential field components (E_z, E_θ, H_z, and H_θ) is continuous at $r = a$. They are expressed as

$$E_z(a - 0, \theta) = E_z(a + 0, \theta), \tag{2.243}$$

$$E_\theta(a - 0, \theta) = E_\theta(a + 0, \theta), \tag{2.244}$$

$$H_z(a - 0, \theta) = H_z(a + 0, \theta), \tag{2.245}$$

$$H_\theta(a - 0, \theta) = H_\theta(a + 0, \theta), \tag{2.246}$$

Substituting the equations for the field components [i.e., Eqs. (2.231)–(2.242)], into the equations for the boundary conditions [i.e., Eqs. (2.243)–(2.246)], we get the matrix equation

$$\begin{pmatrix} J_n & 0 & -K_n & 0 \\ -\dfrac{j\beta}{(u/a)^2}\dfrac{n}{a}J_n & \dfrac{j\omega\mu_0}{u/a}J'_n & -\dfrac{j\beta}{(w/a)^2}\dfrac{n}{a}K_n & \dfrac{j\omega\mu_0}{w/a}K'_n \\ 0 & J_n & 0 & -K_n \\ -\dfrac{j\omega\varepsilon_1}{u/a}J'_n & \dfrac{j\beta}{(u/a)^2}\dfrac{n}{a}J_n & -\dfrac{j\omega\varepsilon_2}{w/a}K'_n & \dfrac{j\beta}{(w/a)^2}\dfrac{n}{a}K_n \end{pmatrix} \begin{pmatrix} A \\ B \\ C \\ D \end{pmatrix} = 0.$$

(2.247)

When the coefficients A, B, C, and D have nontrivial solutions, the determinant of the coefficients of Eq. (2.247) has to be zero. After some mathematical manipulations, we get the characteristic equation

$$\left(\frac{J'_n}{uJ_n} + \frac{K'_n}{wJ_n}\right)\left(\frac{\varepsilon_1}{\varepsilon_2}\frac{J'_n}{uJ_n} + \frac{K'_n}{wJ_n}\right) = n^2\left(\frac{1}{u^2} + \frac{1}{w^2}\right)\left(\frac{\varepsilon_1}{\varepsilon_2}\frac{1}{u^2} + \frac{1}{w^2}\right). \quad (2.248)$$

This is the characteristic equation for the hybrid mode (i.e., $E_z \neq 0$ and $H_z \neq 0$). Since u and w are functions of the effective index n_{eff}, this characteristic equation is also a function of n_{eff}.

For $n = 0$, Eq. (2.248) is reduced to

$$\left(\frac{J'_n}{uJ_n} + \frac{K'_n}{wJ_n}\right) = 0 \quad (2.249)$$

or

$$\left(\frac{\varepsilon_1}{\varepsilon_2}\frac{J'_n}{uJ_n} + \frac{K'_n}{wJ_n}\right) = 0. \quad (2.250)$$

It can be shown that Eqs. (2.249) and (2.250) respectively correspond to the TE mode ($E_z = 0$) and the TM mode ($H_z = 0$).

PROBLEMS

1. Derive the expressions for the power confinement factors (Γ factors) in the core for the TE mode and the TM mode of the three-layer slab optical waveguide shown in Fig. 2.1.

ANSWER

a. TE mode. The principal electric field component is E_y and the other field components are

$$E_x = E_z = H_y = 0, \tag{P2.1}$$

$$H_x = -\frac{\beta}{\omega\mu_0} E_y, \tag{P2.2}$$

$$H_z = \frac{j}{\omega\mu_0} \frac{\partial E_y}{\partial x}. \tag{P2.3}$$

Since the complex Poynting vector **S** is defined as

$$\mathbf{S} = \tfrac{1}{2}\mathbf{E}\times\mathbf{H}^*, \tag{P2.4}$$

the power propagating in the z direction, S_z, is given by

$$\begin{aligned} S_z &= \tfrac{1}{2}(\mathbf{E}\times\mathbf{H}^*)_z \\ &= -\tfrac{1}{2}\int E_y H_x^*\,dx \\ &= \frac{\beta}{2\omega\mu_0}\int |E_y|^2\,dx. \end{aligned} \tag{P2.5}$$

The power confinement factor in the core is thus

$$\Gamma_{\mathrm{TE}} = \frac{(1/2\omega\mu_0)\displaystyle\int_0^W |E_y|^2\,dx}{(1/2\omega\mu_0)\displaystyle\int_{-\infty}^{+\infty} |E_y|^2\,dx}. \tag{P2.6}$$

b. TM mode. The principal electric field component is H_y is and the other field components are.

$$H_x = H_z = E_y = 0, \qquad (P2.7)$$

$$E_x = \frac{\beta}{\omega \varepsilon_0 \varepsilon_r} H_y, \qquad (P2.8)$$

$$E_z = \frac{\beta}{j\omega \varepsilon_0 \varepsilon_r} \frac{\partial H_x}{\partial x}. \qquad (P2.9)$$

The power propagating in the z direction, S_z, is given by

$$S_z = \tfrac{1}{2} \int E_x H_y^* \, dx$$

$$= \frac{\beta}{2\omega\varepsilon_0} \int \frac{1}{\varepsilon_r} |H_y|^2 \, dx. \qquad (P2.10)$$

The power confinement factor in the core is thus

$$\Gamma_{TM} = \frac{(1/2\omega\varepsilon_0) \int_0^W \frac{1}{\varepsilon_r} |H_y|^2 \, dx}{(1/2\omega\varepsilon_0) \int_{-\infty}^{+\infty} \frac{1}{\varepsilon_r} |H_y|^2 \, dx}. \qquad (P2.11)$$

2. The characteristic equations for the LP_{0m} mode, Eq. (2.176), and for the LP_{lm} mode, Eq. (2.183), were derived separately. Show that, when $l = 0$, Eq. (2.176) is included in Eq. (2.183).

ANSWER

Substituting $l = $ into Eq. (2.183), we get

$$\frac{J_0(u)}{uJ_{-1}(u)} = \frac{K_0(w)}{wK_{-1}(w)}. \qquad (P2.12)$$

Since the Bessel functions here have the formulas

$$J_{-n}(z) = (-1)^{-n} J_n(z), \qquad \text{(P2.13)}$$

$$K_{-n}(z) = K_n(z), \qquad \text{(P2.14)}$$

assuming $n = 1$, we get

$$J_{-1}(z) = -J_1(z), \qquad \text{(P2.15)}$$

$$K_{-1}(z) = K_1(z). \qquad \text{(P2.16)}$$

The characteristic equation (2.176) can be derived by substituting Eqs. (P2.15) and (P2.16) into Eq. (P2.12).

REFERENCES

[1] T. Tanaka and Y. Suematsu, "An exact analysis of cylindrical fiber with index distribution by matrix method and its application to focusing fiber," *Trans. IECE Jpn.*, vol. E-59, no. 11, pp. 1–8, 1976.

[2] K. Kawano, *Introduction and Application of Optical Coupling Systems to Optical Devices*, 2nd ed., Gendai Kohgakusha, Tokyo, 1998 (in Japanese).

[3] E. A. Marcatili, "Dielectric rectangular waveguide and directional coupler for integrated optics," *Bell Syst. Tech. J.*, vol. 48, pp. 2071–2102, 1969.

[4] E. Snitzer, "Cylindrical dielectric waveguide modes," *J. Opt. Soc. Am.*, vol. 51, pp. 491–498, 1961.

[5] D. Gloge, "Weakly guiding fibers," *Appl. Opt.*, vol. 10, pp. 2252–2258, 1971.

CHAPTER 3

FINITE-ELEMENT METHODS

Like the finite-difference methods (FDMs) that will be discussed in Chapter 4, the finite-element methods (FEMs) [1–3] are widely utilized numerical methods. The scalar (SC) FEM has several advantages over the fully vectorial (V) FEM. The main ones are that the SC-FEM has no spurious problem and the matrixes in the eigenvalue equation are small and symmetrical These contribute to numerical efficiency. Here, we discuss the use of the SC-FEM in the 2D cross-sectional analysis of optical waveguides.

3.1 VARIATIONAL METHOD

As shown in Fig. 3.1, there are two kinds of ways that can be used to solve optical waveguide problems [2–4]: the variational method and weighted residual method, of which the Galerkin method is representative. Both the variational and the weighted residual methods eventually require that the same matrix eigenvalue equations be solved. This section will focus on a variational method; the next section will discuss a weighted residual method.

In the variational method, the wave equation is not directly solved. Instead, the analysis region is divided into many segments and the variational principle is applied to the sum of the discretized functionals for all segments.

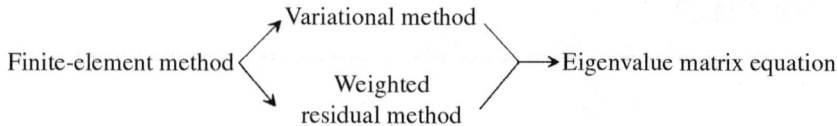

FIGURE 3.1. Analysis based on the finite-element method.

Figure 3.2 shows an analysis region Ω surrounded by a boundary Γ. Here, **n** is the outward-directed unit vector normal to the surface of the analysis region Ω. A variational method for obtaining the effective index n_{eff} is first discussed here by using the scalar wave equation [2,3]

$$\frac{\partial^2 \phi}{\partial x^2} + \frac{\partial^2 \phi}{\partial y^2} + k_0^2(\varepsilon_r - n_{\text{eff}}^2)\phi = 0. \tag{3.1}$$

Multiplying this equation by variation $\delta\phi$ of the function ϕ and integrating the product over the whole analysis region Ω, we get

$$\iint_\Omega \delta\phi \left(\frac{\partial^2 \phi}{\partial x^2} + \frac{\partial^2 \phi}{\partial y^2} \right) dx\, dy + \iint_\Omega \delta\phi\, k_0^2(\varepsilon_r - n_{\text{eff}}^2)\phi dx\, dy = 0. \tag{3.2}$$

Here, applying the partial integration with respect to variables x and y, we get

$$\left[\int_\Gamma \delta\phi \frac{\partial \phi}{\partial x} dy + \int_\Gamma \delta\phi \frac{\partial \phi}{\partial y} dx \right]$$
$$- \iint_\Omega \left(\frac{\partial \delta\phi}{\partial x} \frac{\partial \phi}{\partial x} + \frac{\partial \delta\phi}{\partial y} \frac{\partial \phi}{\partial y} \right) dx\, dy + \iint_\Omega \delta\phi\, k_0^2(\varepsilon_r - n_{\text{eff}}^2)\phi\, dx\, dy = 0,$$

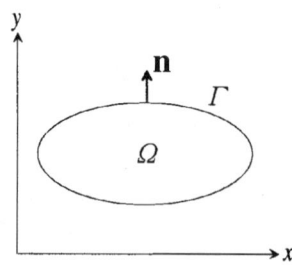

FIGURE 3.2. Analysis region.

where the terms inside the brackets are those whose integration orders with respect to x and y were reduced from 2 to 1 because of the partial integration. Summarizing the second and the third terms, we get

$$\left[\int_\Gamma \delta\phi \frac{\partial \phi}{\partial x} dy + \int_\Gamma \delta\phi \frac{\partial \phi}{\partial y} dx\right]$$
$$- \int\int_\Omega \left\{\left(\frac{\partial \delta\phi}{\partial x}\frac{\partial \phi}{\partial x} + \frac{\partial \delta\phi}{\partial y}\frac{\partial \phi}{\partial y}\right) - \delta\phi\, k_0^2(\varepsilon_r - n_{\text{eff}}^2)\phi\right\} dx\, dy = 0.$$

In addition, substituting the relations

$$\frac{\partial \delta\phi}{\partial x}\frac{\partial \phi}{\partial x} = \frac{\partial \phi}{\partial x}\frac{\partial \delta\phi}{\partial x} = \left(\frac{\partial \phi}{\partial x}\right)\delta\left(\frac{\partial \phi}{\partial x}\right) = \delta\left\{\frac{1}{2}\left(\frac{\partial \phi}{\partial x}\right)^2\right\},$$

$$\frac{\partial \delta\phi}{\partial y}\frac{\partial \phi}{\partial y} = \delta\left\{\frac{1}{2}\left(\frac{\partial \phi}{\partial y}\right)^2\right\},$$

$$\phi\, \delta\phi = \delta\left(\frac{1}{2}\phi^2\right)$$

into the above equation, we get

$$\delta\left[\int_\Gamma \phi \frac{\partial \phi}{\partial x} dy + \int_\Gamma \phi \frac{\partial \phi}{\partial y} dx\right]$$
$$- \delta\frac{1}{2}\left[\int\int_\Omega \left\{\left(\frac{\partial \phi}{\partial x}\right)^2 + \left(\frac{\partial \phi}{\partial y}\right)^2 - k_0^2(\varepsilon_r - n_{\text{eff}}^2)\phi^2\right\} dx\, dy = 0. \quad (3.3)$$

Here, we introduce the following function I:

$$I = \frac{1}{2}\left[\int\int_\Omega \left\{\left(\frac{\partial \phi}{\partial x}\right)^2 + \left(\frac{\partial \phi}{\partial y}\right)^2 - k_0^2(\varepsilon_r - n_{\text{eff}}^2)\phi^2\right\} dx\, dy\right]$$
$$- \left[\int_\Gamma \phi \frac{\partial \phi}{\partial x} dy + \int_\Gamma \phi \frac{\partial \phi}{\partial y} dx\right]. \quad (3.4)$$

Since the line integral calculus term can be rewritten as

$$\int_\Gamma \phi \frac{\partial \phi}{\partial x} dy + \int_\Gamma \phi \frac{\partial \phi}{\partial y} dx = \int_\Gamma \phi \left(\frac{\partial \phi}{\partial x} dy + \frac{\partial \phi}{\partial y} dx \right)$$
$$= \int_\Gamma \phi \frac{\partial \phi}{\partial n} d\Gamma, \qquad (3.5)$$

I can be rewritten as

$$I = \frac{1}{2} \left[\iint_\Omega \left\{ \left(\frac{\partial \phi}{\partial x} \right)^2 + \left(\frac{\partial \phi}{\partial y} \right)^2 - k_0^2 (\varepsilon_r - n_{\text{eff}}^2) \phi^2 \right\} dx\, dy \right] - \left[\int_\Gamma \phi \frac{\partial \phi}{\partial n} d\Gamma \right], \qquad (3.6)$$

where $\partial/\partial n$ is the derivative with respect to the normal vector **n**. Here, I is a function of ϕ, which is a function of x and y. A function of a function is generally called a functional.

Using the functional I, we can rewrite Eq. (3.3) as

$$\delta I = 0, \qquad (3.7)$$

which means that the stationary condition is imposed on the functional I.

We can instead first define the functional I given by Eq. (3.4) or (3.6). Then, imposing the stationary condition on the functional I, we get

$$\delta \phi \iint_\Omega \left\{ \left(\frac{\partial^2 \phi}{\partial x^2} + \frac{\partial^2 \phi}{\partial y^2} \right) + k_0^2 (\varepsilon_r - n_{\text{eff}}^2) \phi \right\} dx\, dy = 0. \qquad (3.8)$$

When Eq. (3.8) holds for an arbitrary variation $\delta \phi$ of the function ϕ, the wave equation (3.1) has to hold. Thus, the wave equation (3.1) is obtained by imposing the stationary condition on the functional I.

This means that *solving the wave equation (3.1) is equivalent to setting the variation δI of the functional I, which can be obtained by Eq. (3.4) or (3.6), to zero, that is, to imposing the stationary condition on the functional I*. This is called the variational principle, and a method for solving problems by using the variational principle is called a variational method. In the Rayleigh–Ritz method, the unknown function is formed by a linear combination of known basis functions that satisfy the boundary conditions, and the variational principle is applied to the functional.

The calculation procedure for the application of the variational method to the FEMs is summarized as follows: The analysis region is first divided into segments, which are called elements, and the functional I_e is calculated for each element e. Then, the total functional I for the whole analysis region is obtained by summing up the functional I_e for all elements:

$$I = \sum_e I_e. \tag{3.9}$$

The final eigenvalue matrix equation is obtained by imposing the stationary condition on the functional I. It should be noted that the difference between an ordinary variational method and an FEM is that the former treats the analysis region as one area and the latter treats the region as the sum of elements.

Since the total functional is a linear combination of the functionals for the elements, the variation δI of the whole system is a sum of variation δI_e of each element e. Thus, Eq. (3.7) can be rewritten as

$$\delta I = \sum_e \delta I_e = 0. \tag{3.10}$$

The functional I_e of an element e surrounded by the boundary Γ_e is given by

$$I_e = \frac{1}{2}\left[\iint_e \left\{\left(\frac{\partial \phi_e}{\partial x}\right)^2 + \left(\frac{\partial \phi_e}{\partial y}\right)^2 - k_0^2(\varepsilon_r - n_{\text{eff}}^2)\phi_e^2\right\} dx\, dy\right]$$
$$- \left[\int_{\Gamma_e} \phi_e \frac{\partial \phi_e}{\partial n} d\Gamma\right]. \tag{3.11}$$

Next, the wave function ϕ_e is expanded as

$$\phi_e = \sum_{i=1}^{M_e} N_i \phi_{ei} = [N_e]^T \{\phi_e\} \tag{3.12}$$

by using the basis function $[N_e]$ in element e, where M_e is the number of nodes in element e and T is the transposing operator for a matrix. Then the basis function $[N_e]$ and the expansion coefficient $\{\phi_e\}$ are expressed as

$$[N_e] = [N_1 \quad N_2 \quad N_3 \quad \cdots \quad N_{M_e}]^T, \tag{3.13}$$
$$\{\phi_e\} = (\phi_1 \quad \phi_2 \quad \phi_3 \quad \cdots \quad \phi_{M_e})^T. \tag{3.14}$$

In the FEMs, a basis function N_i is called a shape function or an interpolation function. As discussed later, the expansion coefficient ϕ_i corresponds to a field component at each node.

Two-dimensional cross-sectional analyses of optical waveguides often use first-order triangular elements having three nodes or second-order triangular elements having six nodes.

We make use of the equation

$$\left(\frac{\partial \phi_e}{\partial x}\right)^2 = \left\{\frac{\partial \phi_e}{\partial x}\right\}^T \left\{\frac{\partial \phi_e}{\partial x}\right\} = \left(\frac{\partial [N_e]^T \{\phi_e\}}{\partial x}\right)^T \frac{\partial [N_e]^T \{\phi_e\}}{\partial x}$$

$$= \frac{\partial \{\phi_e\}^T [N_e]}{\partial x} \frac{\partial [N_e]^T \{\phi_e\}}{\partial x} = \{\phi_e\}^T \frac{\partial [N_e]}{\partial x} \frac{\partial [N_e]^T}{\partial x} \{\phi_e\}, \quad (3.15)$$

where $\partial \phi_e / \partial x$ is a scalar quantity and is rewritten as $(\partial \phi_e / \partial x)^T$. Similarly, we can rewrite

$$\left(\frac{\partial \phi_e}{\partial y}\right)^2 = \{\phi_e\}^T \frac{\partial [N_e]}{\partial y} \frac{\partial [N_e]^T}{\partial y} \{\phi_e\} \quad (3.16)$$

and

$$\phi_e^2 = (\phi_e)^T \phi_e = \{\phi_e\}^T [N_e][N_e]^T \{\phi_e\}, \quad (3.17)$$

where we made use of the fact that ϕ_e is also a scalar quantity.

The functional I_e given by Eq. (3.11) can thus be rewritten as

$$I_e = \frac{1}{2}\{\phi_e\}^T \int\int_e \left(\frac{\partial [N_e]}{\partial x} \frac{\partial [N_e]^T}{\partial x} + \frac{\partial [N_e]}{\partial y} \frac{\partial [N_e]^T}{\partial y}\right.$$

$$\left. - k_0^2(\varepsilon_r - n_{\text{eff}}^2)[N_e][N_e]^T\right) dx\, dy \{\phi_e\} - \left[\int_{\Gamma_e} \phi_e \frac{\partial \phi_e}{\partial n} d\Gamma\right]$$

$$= \frac{1}{2}\{\phi_e\}^T ([A_e] - \lambda^2 [B_e])\{\phi_e\} - \left[\int_{\Gamma_e} \phi_e \frac{\partial \phi_e}{\partial n} d\Gamma\right]. \quad (3.18)$$

3.1 VARIATIONAL METHOD

Here, matrixes $[A_e]$ and $[B_e]$ and the quantity λ^2 are given by

$$[A_e] = \iint_e \left(\frac{\partial [N_e]}{\partial x} \frac{\partial [N_e]^T}{\partial x} + \frac{\partial [N_e]}{\partial y} \frac{\partial [N_e]^T}{\partial y} \right) dx\,dy, \quad (3.19)$$

$$[B_e] = \iint_e [N_e][N_e]^T\, dx\,dy, \quad (3.20)$$

and

$$\lambda^2 = k_0^2(\varepsilon_r - n_{\text{eff}}^2). \quad (3.21)$$

Since the functional I_e given by Eq. (3.18) is obtained for only element e, we have to sum up all the elements to obtain the functional I for the whole analysis region. According to Eq. (3.9),

$$\begin{aligned} I &= \frac{1}{2}\sum_e \{\phi_e\}^T ([A_e] - \lambda^2 [B_e])\{\phi_e\} - \sum_e \left[\int_{\Gamma_e} \phi_e \frac{\partial \phi_e}{\partial n}\, d\Gamma \right] \\ &= \frac{1}{2}\{\phi\}^T ([A] - \lambda^2 [B])\{\phi\} - \sum_e \left[\int_{\Gamma_e} \phi_e \frac{\partial \phi_e}{\partial n}\, d\Gamma \right], \end{aligned} \quad (3.22)$$

where

$$\{\phi\} = \sum_e \{\phi_e\}, \quad (3.23)$$

$$[A] = \sum_e [A_e], \quad (3.24)$$

$$[B] = \sum_e [B_e]. \quad (3.25)$$

Now, consider the second term of Eq. (3.22), which is the line integral calculus term and can be expressed as follows:

$$-\sum_e \left[\int_{\Gamma_e} \phi_e \frac{\partial \phi_e}{\partial n}\, d\Gamma \right] = -\sum_e \left(\int_{\Gamma_{e+1}} \phi_{e+1} \frac{\partial \phi_{e+1}}{\partial n}\, d\Gamma - \int_{\Gamma_e} \phi_e \frac{\partial \phi_e}{\partial n}\, d\Gamma \right).$$

Here, we assume that the wave function ϕ_e and its derivative with respect to the normal to the surface of element e, $\partial \phi_e/\partial n$, are continuous at the boundaries with neighboring elements. Since, under this assumption, the line integral calculus terms inside the analysis region are canceled out, the line integral calculus term of the periphery of the whole analysis region remains, which will be discussed in Section 3.4.

Thus, the second term of Eq. (3.22) can be reduced to

$$-\left(\oint_\Gamma \phi \frac{\partial \phi}{\partial n} d\Gamma\right). \tag{3.26}$$

Finally, the functional I for the whole analysis region given by Eq. (3.22) is obtained as

$$I = \tfrac{1}{2}\{\phi\}^T([A] - \lambda^2[B])\{\phi\} - \left(\oint_\Gamma \phi \frac{\partial \phi}{\partial n} d\Gamma\right). \tag{3.27}$$

Next, we impose on the functional I at the boundaries the Dirichlet condition

$$\phi = 0 \tag{3.28}$$

or the Neumann condition

$$\frac{\partial \phi}{\partial n} = 0. \tag{3.29}$$

Using these boundary conditions, we can reduce the functional I to the simple form

$$\begin{aligned} I &= \frac{1}{2}\sum_e \{\phi_e\}^T([A_e] - \lambda^2[B_e])\{\phi_e\} \\ &= \frac{1}{2}\{\phi\}^T([A] - \lambda^2[B])\{\phi\}. \end{aligned} \tag{3.30}$$

Now that we have obtained the functional I, the next step is to impose the variational principle on it. Although having a total of only two nodes is impossible for 2D cross-sectional optical waveguides, this case will be used here for ease of understanding.

The wave function vector $\{\phi\}$ and matrixes $[S]$, $[A]$, and $[B]$ for the eigenvalue λ^2 are expressed as

$$\{\phi\} = \begin{pmatrix} \phi_1 \\ \phi_2 \end{pmatrix} \tag{3.31}$$

and

$$[S] = [A] - \lambda^2[B] = \begin{pmatrix} S_{11} & S_{12} \\ S_{12} & S_{22} \end{pmatrix}. \qquad (3.32)$$

Here, we make use of the symmetry for $[S]$. [The symmetry of the matrix, $[S] = [S]^T$, will be discussed later.] Thus, we get

$$I = \tfrac{1}{2}\{\phi\}^T[S]\{\phi\} \qquad (3.33)$$

$$= \tfrac{1}{2}(\phi_1 \phi_2)\begin{pmatrix} S_{11} & S_{12} \\ S_{12} & S_{22} \end{pmatrix}\begin{pmatrix} \phi_1 \\ \phi_2 \end{pmatrix}$$

$$= \tfrac{1}{2}(\phi_1 \phi_2)\begin{pmatrix} S_{11}\phi_1 + S_{12}\phi_2 \\ S_{12}\phi_1 + S_{22}\phi_2 \end{pmatrix}$$

$$= \tfrac{1}{2}(S_{11}\phi_1^2 + 2S_{12}\phi_1\phi_2 + S_{22}\phi_2^2). \qquad (3.34)$$

The derivatives of Eq. (3.34) with respect to ϕ_1 and ϕ_2 are

$$\frac{\partial I}{\partial \phi_1} = S_{11}\phi_1 + S_{12}\phi_2, \qquad (3.35)$$

$$\frac{\partial I}{\partial \phi_2} = S_{12}\phi_1 + S_{22}\phi_2. \qquad (3.36)$$

And the matrix expression for these derivatives is

$$\frac{\partial I}{\partial \begin{pmatrix} \phi_1 \\ \phi_2 \end{pmatrix}} = \begin{pmatrix} S_{11} & S_{12} \\ S_{12} & S_{22} \end{pmatrix}\begin{pmatrix} \phi_1 \\ \phi_2 \end{pmatrix} \qquad (3.37)$$

or

$$\frac{\partial I}{\partial \{\phi\}} = ([A] - \lambda^2[B])\{\phi\}. \qquad (3.38)$$

This is the same equation we get for general cases. Since solving the wave equation by using the FEM with the variational principle is the imposition of the stationary condition on the functional I as

$$\frac{\partial I}{\partial \{\phi\}} = 0, \tag{3.39}$$

it is equivalent to solving the eigenvalue matrix equation

$$([A] - \lambda^2 [B])\{\phi\} = 0. \tag{3.40}$$

3.2 GALERKIN METHOD

Since the functional is used to solve the problems, we have to find it when we use the FEM with the variational principle. On the other hand, the partial differential equations governing almost all the physical phenomena that we encounter are already known. This is certainly true with regard to the wave equations for the electromagnetic fields and for the Schrödinger equation, solutions of which are the targets of this book. The weighted residual methods, especially the Galerkin method, is quite powerful for solving them. The Galerkin method is widely used not only in the FEM but also in methods of microwave analyses, such as the spectral domain approach (SDA) [5, 6].

Since the wave function ϕ is a true solution for the wave equation (3.1), the right-hand term of Eq. (3.1) is definitely zero. The true wave function ϕ, however, cannot actually be known; we can obtain only an approximate wave function $\bar{\phi}$. When the true wave function ϕ in Eq. (3.1) is replaced by the approximate one, the right-hand term does not become zero but generates the error R, which is called the error residual:

$$\frac{\partial^2 \bar{\phi}}{\partial x^2} + \frac{\partial^2 \bar{\phi}}{\partial y^2} + k_0^2(\varepsilon_r - n_{\text{eff}}^2)\bar{\phi} = R. \tag{3.41}$$

It is natural to think that the difference between $\bar{\phi}$ and ϕ can be decreased by *averagely* setting the error residual R equal to zero in the whole analysis region. As is well known, the electromagnetic fields concentrate mostly in the core where the refractive index is higher than in the cladding. Thus, some weighting should be used when setting R equal to zero. Introducing the weight function ψ, we get

$$\iint \psi R \, dx \, dy = 0. \tag{3.42}$$

Rewriting the error residual R explicitly, we get

$$\iint \psi \left\{ \frac{\partial^2 \bar{\phi}}{\partial x^2} + \frac{\partial^2 \bar{\phi}}{\partial y^2} + k_0^2(\varepsilon_r - n_{\text{eff}}^2)\bar{\phi} \right\} dx\, dy = 0. \qquad (3.43)$$

The procedure discussed above is the weighted residual method.
Partially integrating Eq. (3.43) with respect to x and y, we get

$$\left[\int \psi \frac{\partial \bar{\phi}}{\partial x} dy + \int \psi \frac{\partial \bar{\phi}}{\partial y} dx \right] - \iint \left(\frac{\partial \psi}{\partial x} \frac{\partial \bar{\phi}}{\partial x} + \frac{\partial \psi}{\partial y} \frac{\partial \bar{\phi}}{\partial y} \right) dx\, dy$$

$$+ \iint \psi k_0^2(\varepsilon_r - n_{\text{eff}}^2)\bar{\phi}\, dx\, dy = 0, \qquad (3.44)$$

which can be rewritten as

$$\left[\int_\Gamma \psi \frac{\partial \bar{\phi}}{\partial n} d\Gamma \right] - \iint \left(\frac{\partial \psi}{\partial x} \frac{\partial \bar{\phi}}{\partial x} + \frac{\partial \psi}{\partial y} \frac{\partial \bar{\phi}}{\partial y} \right) dx\, dy$$

$$+ \iint \psi k_0^2(\varepsilon_r - n_{\text{eff}}^2)\bar{\phi}\, dx\, dy = 0. \qquad (3.45)$$

Here, the relation given by Eq. (3.5) was used, and $\int_\Gamma d\Gamma$ and $\partial/\partial n$ are respectively the line integral calculus at the boundary Γ and the derivative with respect to the normal vector $\mathbf{n}$. The rank of the integral calculus of the first term inside the square brackets in Eq. (3.45) is decreased by 1 as a result of the partial integration. The rank of the derivatives of the second term is also decreased from 2 to 1, and these derivatives are called the weak forms.

The weighted residual method, in which both the approximate wave function $\bar{\phi}$ and the weight function ψ are expanded by the same basis functions, is called the Galerkin method. In using the FEM, we first divide the analysis region into many elements, then apply the Galerkin method to each element, then sum up the contributions of all the elements. The expansion coefficients obtained as an eigenvector correspond to the fields at nodes in the analysis region.

Since the calculation procedure will be discussed in detail later, here we simply summarize it. The equation for element e in the divided elements is expressed from Eq. (3.45) as

$$\left[\int_{\Gamma_e} \psi_e \frac{\partial \bar{\phi}_e}{\partial n} d\Gamma\right] - \iint_e \left(\frac{\partial \psi_e}{\partial x}\frac{\partial \bar{\phi}_e}{\partial x} + \frac{\partial \psi_e}{\partial y}\frac{\partial \bar{\phi}_e}{\partial y}\right) dx\, dy$$

$$+ \iint_e \psi_e k_0^2(\varepsilon_r - n_{\text{eff}}^2)\bar{\phi}_e\, dx\, dy = 0. \qquad (3.46)$$

Here, it should be noted that $\bar{\phi}_e$ and ψ_e in element e are expanded by using the same basis functions:

$$\bar{\phi}_e = \sum_i^{M_e} \phi_{ei} N_i = [N_e]^T \{\phi_e\}, \qquad (3.47)$$

$$\psi_e = \sum_i^{M_e} \phi_{ei} N_i = [N_e]^T \{\phi_e\}, \qquad (3.48)$$

where M_e is the number of nodes in e. The $[N_e]$ and $\{\phi_e\}$ were defined in Eqs. (3.13) and (3.14). The basis function N_i and the expansion coefficient ϕ_{ei} correspond to the shape function and the field component.

Substituting Eqs. (3.47) and (3.48) into Eq. (3.46), we get

$$\{\phi_e\}^T \iint_e \left\{ -\frac{\partial[N_e]}{\partial x}\frac{\partial[N_e]^T}{\partial x} - \frac{\partial[N_e]}{\partial y}\frac{\partial[N_e]^T}{\partial y} \right.$$

$$\left. + k_0^2(\varepsilon_r - n_{\text{eff}}^2)[N_e][N_e]^T \right\} dx\, dy \{\phi_e\} + \left[\int_{\Gamma_e} \bar{\phi}_e \frac{\partial \bar{\phi}_e}{\partial n} d\Gamma\right] = 0$$

$$(3.49)$$

for element e. Applying the definitions for $[A_e]$, $[B_e]$, and λ^2 shown in Eqs. (3.19) to (3.21), we can rewrite the above equation as

$$\{\phi_e\}^T(-[A_e] + \lambda^2[B_e])\{\phi_e\} + \left[\int_{\Gamma_e} \bar{\phi}_e \frac{\partial \bar{\phi}_e}{\partial n} d\Gamma\right] = 0$$

or

$$\{\phi_e\}^T([A_e] - \lambda^2[B_e])\{\phi_e\} - \left[\int_{\Gamma_e} \bar{\phi}_e \frac{\partial \bar{\phi}_e}{\partial n} d\Gamma\right] = 0. \qquad (3.50)$$

Since Eq. (3.50) gives the contribution of only element e, it is necessary to sum up the contributions of all the elements in the analysis region. Thus, we get

$$\sum_e \{\phi_e\}^T ([A_e] - \lambda^2 [B_e])\{\phi_e\} - \sum_e \left[\int_{\Gamma_e} \bar{\phi}_e \frac{\partial \bar{\phi}_e}{\partial n} d\Gamma \right] = 0$$

or

$$\{\phi\}^T ([A] - \lambda^2 [B])\{\phi\} - \sum_e \left[\int_{\Gamma_e} \bar{\phi}_e \frac{\partial \bar{\phi}_e}{\partial n} d\Gamma \right] = 0, \quad (3.51)$$

where the definitions

$$\{\phi\} = \sum_e \{\phi_e\}, \quad [A] = \sum_e [A_e], \quad [B] = \sum_e [B_e] \quad (3.52)$$

in Eqs. (3.23)–(3.25) were used.

With respect to the second term in Eq. (3.51), we assume that the wave function ϕ_e and its normal derivative $\partial \phi_e / \partial n$ are continuous at the boundaries between elements. This assumption cancels out the line integral calculus terms, so the second term in Eq. (3.51) can be reduced to

$$-\left(\oint_\Gamma \bar{\phi} \frac{\partial \bar{\phi}}{\partial n} d\Gamma \right). \quad (3.53)$$

Substituting Eq. (3.53) into (3.51), we get

$$\{\phi\}^T ([A] - \lambda^2 [B])\{\phi\} - \left(\oint_\Gamma \bar{\phi} \frac{\partial \bar{\phi}}{\partial n} d\Gamma \right) = 0. \quad (3.54)$$

When the Dirichlet condition or the Neumann condition is the boundary condition, the second term in Eq. (3.51) becomes zero and Eq. (3.54) is simplified to

$$\{\phi\}^T ([A] - \lambda^2 [B])\{\phi\} = 0. \quad (3.55)$$

We finally get the following eigenvalue matrix equation to be solved:

$$([A] - \lambda^2[B])\{\phi\} = 0. \tag{3.56}$$

Comparing Eq. (3.56) with Eq. (3.40), one can see that the eigenvalue matrix equation for the Galerkin method is identical to that for a variational method.

3.3 AREA COORDINATES AND TRIANGULAR ELEMENTS

We have roughly discussed the calculation procedures for the variational method and Galerkin method. Before describing the detailed formulations of the global matrixes for the eigenvalue matrix equations, we have to investigate the *elements*, which are indispensable when we divide the analysis region into segments.

In the analysis of 2D cross-sectional structures, triangular elements using polynomials are generally used to approximate field distributions. The concept of polynomial approximations is illustrated in Fig. 3.3. In Fig. 3.3a, the true wave function is approximated by a linear function, and, in Fig. 3.3b, it is approximated by a quadratic function.

Although the higher order polynomial approximations can bring about more accurate results, they result in a larger number of nodes at which optical fields are defined. In addition, when there are more nodes, we need to use a more complicated mathematical analysis and more computer memory. Discussions in this book are therefore limited to two widely utilized triangular elements: the first-order triangular element, which requires three nodes, and the second-order triangular element, which requires six nodes.

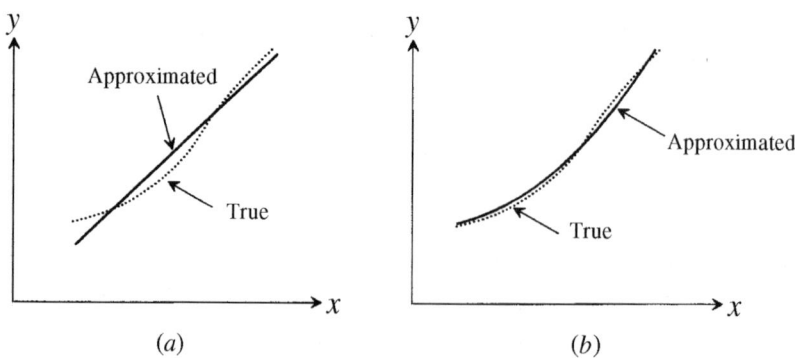

FIGURE 3.3. Approximations by polynomial functions: (*a*) linear function; (*b*) quadratic function.

3.3 AREA COORDINATES AND TRIANGULAR ELEMENTS

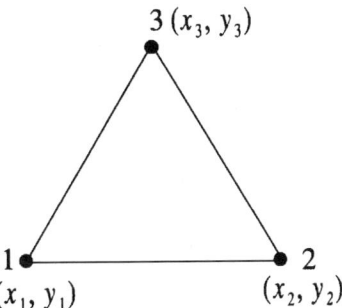

FIGURE 3.4. First-order triangular element.

3.3.1 First-Order Triangular Elements

Figure 3.4 shows a first-order triangular element. In this element, the wave function $\phi(x, y)$ at an arbitrary coordinate (x, y) inside the element is expanded using shape functions N_1, N_2, and N_3 (with fields ϕ_1, ϕ_2, and ϕ_3) at the vertexes on which nodes are placed:

$$\phi(x, y) = N_1\phi_1 + N_2\phi_2 + N_3\phi_3$$
$$= [N]^T\{\phi\}, \tag{3.57}$$

where $[N] = [N_1 \ N_2 \ N_3]^T$ and $\{\phi\} = (\phi_1 \ \phi_2 \ \phi_3)^T$. The shape functions $[N]$ and the field vectors $\{\phi\}$ respectively correspond to the basis functions and the expansion coefficients. The coordinates of nodes 1, 2, and 3 are respectively (x_1, y_1), (x_2, y_2) and (x_3, y_3).

In determining the explicit forms of shape functions N_1, N_2, and N_3, the use of area coordinates is convenient.

Figure 3.5 shows a triangle we use here for discussing the area coordinates. An arbitrary coordinate in this triangle is denoted by $p(x, y)$. Figure 3.6 shows another triangle formed by nodes denoted 1, 2, and 3. As is well known, the area S_{123} of the triangle is given by

$$S_{123} = \tfrac{1}{2}|\mathbf{A} \times \mathbf{B}|. \tag{3.58}$$

When we assume that **i**, **j**, and **k** are unit vectors in the x, y, and z directions, the vector **A** from node 1 to node 2 and the vector **B** from node 1 to node 3 are expressed as

$$\mathbf{A} = (x_2 - x_1)\mathbf{i} + (y_2 - y_1)\mathbf{j}, \qquad \mathbf{B} = (x_3 - x_1)\mathbf{i} + (y_3 - y_1)\mathbf{j}. \tag{3.59}$$

74 FINITE-ELEMENT METHODS

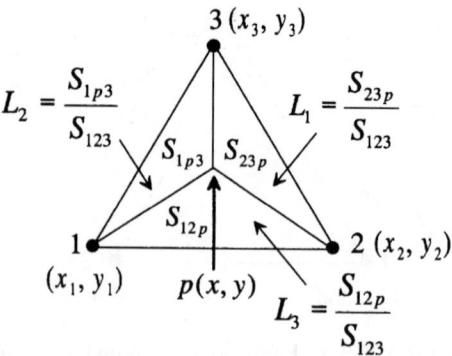

FIGURE 3.5. Area coordinates.

Using these expressions, we get

$$\mathbf{A} \times \mathbf{B} = \begin{vmatrix} \mathbf{i} & \mathbf{j} & \mathbf{k} \\ x_2 - x_1 & y_2 - y_1 & 0 \\ x_3 - x_1 & y_3 - y_1 & 0 \end{vmatrix}$$

$$= \begin{vmatrix} x_2 - x_1 & y_2 - y_1 \\ x_3 - x_1 & y_3 - y_1 \end{vmatrix} \mathbf{k}. \qquad (3.60)$$

Thus, we can obtain the area S_{123} of the triangle 123 as

$$S_{123} = \frac{1}{2} \begin{vmatrix} x_2 - x_1 & y_2 - y_1 \\ x_3 - x_1 & y_3 - y_1 \end{vmatrix}. \qquad (3.61)$$

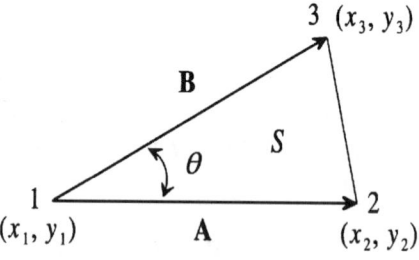

FIGURE 3.6. Area of a triangle.

We similarly get area S_{23p} of triangle $23p$ formed by nodes 2 and 3 and point p; area S_{1p3} of triangle $1p3$ formed by node 1, point p, and node 3; and area S_{12p} of triangle $12p$ formed by nodes 1 and 2 and point p:

$$S_{23p} = \frac{1}{2}\begin{vmatrix} x_2 - x & y_2 - y \\ x_3 - x & y_3 - y \end{vmatrix}, \qquad (3.62)$$

$$S_{1p3} = \frac{1}{2}\begin{vmatrix} x_3 - x & y_3 - y \\ x_1 - x & y_1 - y \end{vmatrix}, \qquad (3.63)$$

$$S_{12p} = \frac{1}{2}\begin{vmatrix} x_2 - x & y_2 - y \\ x_1 - x & y_1 - y \end{vmatrix}. \qquad (3.64)$$

Area coordinate L_i can be defined as *the ratio of the area of the triangle formed by point p and the side opposite node i to the whole area of the triangle*. In Fig. 3.5, the area coordinate L_1, which is related to node 1, is defined as the ratio of the area of triangle $23p$ to the area of triangle 123. Similarly, the area coordinate L_2, related to node 2, is defined as the ratio of the area of triangle $1p3$ to the area of triangle 123. And area coordinate L_3, related to node 3, is defined as the ratio of the area of triangle $12p$ to the area of triangle 123. The explicit expressions for L_1, L_2, and L_3 are

$$L_1 = \frac{S_{23p}}{S_{123}} = \frac{\begin{vmatrix} x_2 - x & y_2 - y \\ x_3 - x & y_3 - y \end{vmatrix}}{\begin{vmatrix} x_2 - x_1 & y_2 - y_1 \\ x_3 - x_1 & y_3 - y_1 \end{vmatrix}}, \qquad (3.65)$$

$$L_2 = \frac{S_{1p3}}{S_{123}} = \frac{\begin{vmatrix} x_3 - x & y_3 - y \\ x_1 - x & y_1 - y \end{vmatrix}}{\begin{vmatrix} x_2 - x_1 & y_2 - y_1 \\ x_3 - x_1 & y_3 - y_1 \end{vmatrix}}, \qquad (3.66)$$

$$L_3 = \frac{S_{12p}}{S_{123}} = \frac{\begin{vmatrix} x_2 - x & y_2 - y \\ x_1 - x & y_1 - y \end{vmatrix}}{\begin{vmatrix} x_2 - x_1 & y_2 - y_1 \\ x_3 - x_1 & y_3 - y_1 \end{vmatrix}}, \qquad (3.67)$$

Generally, since the value of the determinant for a matrix is invariant under transposition, we get the relations

$$\begin{vmatrix} x_2 - x_1 & y_2 - y_1 \\ x_3 - x_1 & y_3 - y_1 \end{vmatrix} = \begin{vmatrix} x_1 & y_1 & 1 \\ x_2 & y_2 & 1 \\ x_3 & y_3 & 1 \end{vmatrix} = \begin{vmatrix} x_1 & x_2 & x_3 \\ y_1 & y_2 & y_3 \\ 1 & 1 & 1 \end{vmatrix} (= 2S_{123}). \qquad (3.68)$$

We similarly get the following relations for the other determinants:

$$\begin{vmatrix} x_2 - x & y_2 - y \\ x_3 - x & y_3 - y \end{vmatrix} = \begin{vmatrix} x & y & 1 \\ x_2 & y_2 & 1 \\ x_3 & y_3 & 1 \end{vmatrix} = \begin{vmatrix} x & x_2 & x_3 \\ y & y_2 & y_3 \\ 1 & 1 & 1 \end{vmatrix},$$

$$\begin{vmatrix} x_3 - x & y_3 - y \\ x_1 - x & y_1 - y \end{vmatrix} = \begin{vmatrix} x & y & 1 \\ x_3 & y_3 & 1 \\ x_1 & y_1 & 1 \end{vmatrix} = \begin{vmatrix} x & x_3 & x_1 \\ y & y_3 & y_1 \\ 1 & 1 & 1 \end{vmatrix} = \begin{vmatrix} x_1 & x & x_3 \\ y_1 & y & y_3 \\ 1 & 1 & 1 \end{vmatrix},$$

$$\begin{vmatrix} x_2 - x & y_2 - y \\ x_1 - x & y_1 - y \end{vmatrix} = \begin{vmatrix} x & y & 1 \\ x_2 & y_2 & 1 \\ x_1 & y_1 & 1 \end{vmatrix} = \begin{vmatrix} x & x_2 & x_1 \\ y & y_2 & y_1 \\ 1 & 1 & 1 \end{vmatrix} = \begin{vmatrix} x_1 & x_2 & x \\ y_1 & y_2 & y \\ 1 & 1 & 1 \end{vmatrix}.$$

Rewriting Eqs. (3.65)–(3.67) by using these relations, we get

$$L_1 = \begin{vmatrix} x & x_2 & x_3 \\ y & y_2 & y_3 \\ 1 & 1 & 1 \end{vmatrix} \bigg/ \begin{vmatrix} x_1 & x_2 & x_3 \\ y_1 & y_2 & y_3 \\ 1 & 1 & 1 \end{vmatrix}, \tag{3.69}$$

$$L_2 = \begin{vmatrix} x_1 & x & x_3 \\ y_1 & y & y_3 \\ 1 & 1 & 1 \end{vmatrix} \bigg/ \begin{vmatrix} x_1 & x_2 & x_3 \\ y_1 & y_2 & y_3 \\ 1 & 1 & 1 \end{vmatrix}, \tag{3.70}$$

$$L_3 = \begin{vmatrix} x_1 & x_2 & x \\ y_1 & y_2 & y \\ 1 & 1 & 1 \end{vmatrix} \bigg/ \begin{vmatrix} x_1 & x_2 & x_3 \\ y_1 & y_2 & y_3 \\ 1 & 1 & 1 \end{vmatrix}, \tag{3.71}$$

As shown in these equations, area coordinates L_1, L_2, and L_3 are linear functions of x and y. According to Cramer's formula, Eqs. (3.69)–(3.71) mean that L_1, L_2, and L_3 are roots of the following algebraic matrix equations:

$$\begin{pmatrix} x \\ y \\ 1 \end{pmatrix} = \begin{pmatrix} x_1 & x_2 & x_3 \\ y_1 & y_2 & y_3 \\ 1 & 1 & 1 \end{pmatrix} \begin{pmatrix} L_1 \\ L_2 \\ L_3 \end{pmatrix}, \tag{3.72}$$

where it can be found from the bottom row that the sum of L_1, L_2, and L_3 is 1.

3.3 AREA COORDINATES AND TRIANGULAR ELEMENTS

Now, we return our attention to the shape functions in Eq. (3.57). Since here we are interested in a first-order triangular element, we approximate shape functions N_1, N_2, and N_3 by the following first-order plane functions:

$$N_1 = a_1 x + b_1 y + c_1, \tag{3.73}$$
$$N_2 = a_2 x + b_2 y + c_2, \tag{3.74}$$
$$N_3 = a_3 x + b_3 y + c_3, \tag{3.75}$$

for which the matrix expression is

$$\begin{pmatrix} N_1 \\ N_2 \\ N_3 \end{pmatrix} = \begin{pmatrix} a_1 & b_1 & c_1 \\ a_2 & b_2 & c_2 \\ a_3 & b_3 & c_3 \end{pmatrix} \begin{pmatrix} x \\ y \\ 1 \end{pmatrix}. \tag{3.76}$$

The conditions to be imposed on shape function N_1 are the following:

When $p(x, y)$ coincides with node 1 (x_1, y_1): $N_1 = a_1 x_1 + b_1 y_1 + c_1 = 1$, \hfill (3.77)

When $p(x, y)$ coincides with node 2 (x_2, y_2): $N_1 = a_1 x_2 + b_1 y_2 + c_1 = 0$, \hfill (3.78)

When $p(x, y)$ coincides with node 3 (x_3, y_3): $N_1 = a_1 x_3 + b_1 y_3 + c_1 = 0$. \hfill (3.79)

The matrix expression for Eqs. (3.77)–(3.79) is

$$\begin{pmatrix} x_1 & y_1 & 1 \\ x_2 & y_2 & 1 \\ x_3 & y_3 & 1 \end{pmatrix} \begin{pmatrix} a_1 \\ b_1 \\ c_1 \end{pmatrix} = \begin{pmatrix} 1 \\ 0 \\ 0 \end{pmatrix}. \tag{3.80}$$

Imposing the same conditions on N_2 and N_3, we get

$$\begin{pmatrix} x_1 & y_1 & 1 \\ x_2 & y_2 & 1 \\ x_3 & y_3 & 1 \end{pmatrix} \begin{pmatrix} a_2 \\ b_2 \\ c_2 \end{pmatrix} = \begin{pmatrix} 0 \\ 1 \\ 0 \end{pmatrix}, \tag{3.81}$$

$$\begin{pmatrix} x_1 & y_1 & 1 \\ x_2 & y_2 & 1 \\ x_3 & y_3 & 1 \end{pmatrix} \begin{pmatrix} a_3 \\ b_3 \\ c_3 \end{pmatrix} = \begin{pmatrix} 0 \\ 0 \\ 1 \end{pmatrix}. \tag{3.82}$$

Equations (3.80)–(3.82) can be rewritten as

$$\begin{pmatrix} x_1 & y_1 & 1 \\ x_2 & y_2 & 1 \\ x_3 & y_3 & 1 \end{pmatrix} \begin{pmatrix} a_1 & a_2 & a_3 \\ b_1 & b_2 & b_3 \\ c_1 & c_2 & c_3 \end{pmatrix} = \begin{pmatrix} 1 & 0 & 0 \\ 0 & 1 & 0 \\ 0 & 0 & 1 \end{pmatrix}. \quad (3.83)$$

Using the definitions

$$[X] = \begin{pmatrix} x_1 & y_1 & 1 \\ x_2 & y_2 & 1 \\ x_3 & y_3 & 1 \end{pmatrix}, \quad [A] = \begin{pmatrix} a_1 & a_2 & a_3 \\ b_1 & b_2 & b_3 \\ c_1 & c_2 & c_3 \end{pmatrix}, \quad [I] = \begin{pmatrix} 1 & 0 & 0 \\ 0 & 1 & 0 \\ 0 & 0 & 1 \end{pmatrix},$$
$$(3.84)$$

we can rewrite Eq. (3.83) as

$$[X][A] = [I]. \quad (3.85)$$

Taking the transposition T of Eq. (3.85), we get

$$[A]^T[X]^T = [I]. \quad (3.86)$$

Thus, matrix $[X]^T$ is found to be an inverse matrix of $[A]^T$. Multiplying Eq. (3.76) by matrix $[X]^T$, we get

$$[X]^T \begin{pmatrix} N_1 \\ N_2 \\ N_3 \end{pmatrix} = [X]^T[A]^T \begin{pmatrix} x \\ y \\ 1 \end{pmatrix} = [I] \begin{pmatrix} x \\ y \\ 1 \end{pmatrix} = \begin{pmatrix} x \\ y \\ 1 \end{pmatrix}.$$

This results in

$$\begin{pmatrix} x \\ y \\ 1 \end{pmatrix} = [X]^T \begin{pmatrix} N_1 \\ N_2 \\ N_3 \end{pmatrix} = \begin{pmatrix} x_1 & x_2 & x_3 \\ y_1 & y_2 & y_3 \\ 1 & 1 & 1 \end{pmatrix} \begin{pmatrix} N_1 \\ N_2 \\ N_3 \end{pmatrix}. \quad (3.87)$$

Comparing Eq. (3.72) with Eq. (3.87), we find the following important relations between shape functions N_1, N_2, and N_3 and area coordinates L_1, L_2, and L_3:

$$N_1 = L_1, \quad N_2 = L_2, \quad N_3 = L_3. \quad (3.88)$$

3.3 AREA COORDINATES AND TRIANGULAR ELEMENTS

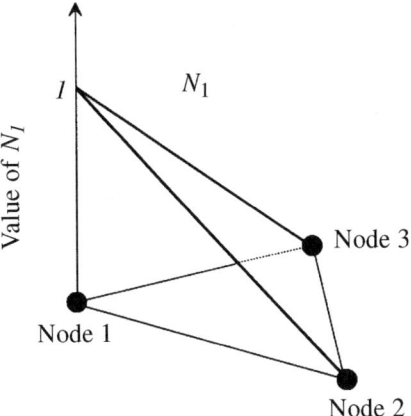

FIGURE 3.7. Shape function N_1 for the first-order triangular element.

Figure 3.7 shows N_1 as an example. As shown in this figure, N_1 is a plane function whose value is 1 at node 1. Similarly, N_2 and N_3 are respectively plane functions whose values are 1 at nodes 2 and 3.

3.3.2 Second-Order Triangular Elements

Figure 3.8 shows a second-order triangular element. As in the first-order triangular element shown in Fig. 3.4, nodes 1, 2, and 3 are positioned at the vertexes of the triangle. The second-order triangular element has three additional nodes set midway between the pairs of vertexes.

The wave function $\phi(x, y)$ at an arbitrary coordinate (x, y) is expanded using six shape functions and the values of wave functions $\phi_1, \ldots, \phi_6$ at the nodes as

$$\phi(x, y) = \sum_{i=1}^{6} N_i \phi_i = [N]^T \{\phi\}, \tag{3.89}$$

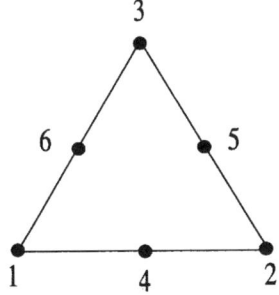

FIGURE 3.8. Second-order triangular element.

where $[N] = [N_1 \; N_2 \; N_3 \; N_4 \; N_5 \; N_6]^T$ and $\{\phi\} = (\phi_1 \; \phi_2 \; \phi_3 \; \phi_4 \; \phi_5 \; \phi_6)^T$. The coordinates of nodes 1–6 are respectively $(x_1, y_1), \ldots, (x_6, y_6)$.

As mentioned before, the first-order shape function is a linear function of x and y. The second-order shape functions should be quadratic polynomial functions of x and y. The area coordinates L_1, L_2, and L_3 are also linear functions of x and y, so the second-order shape functions must be quadratic functions of the area coordinates.

A. Shape Function N_1 Shape function N_1 is 1 at node 1 and is 0 at all other nodes. Since $L_1 = 1$ and $L_2 = L_3 = 0$ at node 1, N_1 should be a quadratic function of only area coordinate L_1. Thus,

$$N_1 = aL_1^2 + bL_1 + c. \tag{3.90}$$

The following relations should hold:

At node 1 ($L_1 = 1$): $N_1 = a + b + c = 1.$ (3.91)

At node 6 ($L_1 = 0.5$): $N_1 = 0.25a + 0.5b + c = 0.$ (3.92)

At node 3 ($L_1 = 0$): $N_1 = c = 0.$ (3.93)

From Eqs. (3.91)–(3.93), we obtain

$$a = 2, \quad b = -1, \quad c = 0. \tag{3.94}$$

Thus, we get

$$N_1 = 2L_1^2 - L_1 = L_1(2L_1 - 1). \tag{3.95}$$

B. Shape Function N_2 Shape function N_2 is 1 at node 2 and is 0 at all other nodes. Since $L_2 = 1$ and $L_1 = L_3 = 0$ at node 2, N_2 should be a quadratic function of only area coordinate L_2. Thus,

$$N_2 = aL_2^2 + bL_2 + c. \tag{3.96}$$

3.3 AREA COORDINATES AND TRIANGULAR ELEMENTS

The following relations should hold:

At node 2 ($L_2 = 1$): $\quad N_2 = a + b + c = 1.$ (3.97)

At node 4 ($L_2 = 0.5$): $\quad N_2 = 0.25a + 0.5b + c = 0.$ (3.98)

At node 1 ($L_2 = 0$): $\quad N_2 = c = 0.$ (3.99)

From Eqs. (3.97)–(3.99), we obtain

$$a = 2, \quad b = -1, \quad c = 0. \tag{3.100}$$

Thus, we get

$$N_2 = L_2(2L_2 - 1). \tag{3.101}$$

C. Shape Function N_3 Shape function N_3 is 1 at node 3 and is 0 at all other nodes. Since $L_3 = 1$ and $L_1 = L_2 = 0$ at node 3, N_3 should be a quadratic function of only area coordinate L_3. Thus,

$$N_3 = aL_3^2 + bL_3 + c. \tag{3.102}$$

The following relations should hold:

At node 3 ($L_3 = 1$): $\quad N_3 = a + b + c = 1.$ (3.103)

At node 6 ($L_3 = 0.5$): $\quad N_3 = 0.25a + 0.5b + c = 0.$ (3.104)

At node 1 ($L_3 = 0$): $\quad N_3 = c = 0.$ (3.105)

From Eqs. (3.103)–(3.105), we obtain

$$a = 2, \quad b = -1, \quad c = 0. \tag{3.106}$$

Thus, we get

$$N_3 = L_3(2L_3 - 1). \tag{3.107}$$

D. Shape Function N_4 Shape function N_4 is 1 at node 4 and is 0 at all other nodes. Since $L_1 = L_2 = 0.5$ and $L_3 = 0$ at node 4, N_4 should be a quadratic function of area coordinates L_1 and L_2. Thus,

$$N_4 = aL_1^2 + bL_2^2 + cL_1L_2 + dL_1 + eL_2 + f. \quad (3.108)$$

The following relations should hold:

At node 4 ($L_1 = L_2 = 0.5$): $\quad N_4 = 0.25a + 0.25b + 0.25c$
$\qquad\qquad\qquad\qquad\qquad\qquad + 0.5d + 0.5e + f = 1.\quad$ (3.109)
At node 1 ($L_1 = 1, L_2 = 0$): $\quad N_4 = a + d + f = 0.\quad$ (3.110)
At node 2 ($L_1 = 0, L_2 = 1$): $\quad N_4 = b + e + f = 0.\quad$ (3.111)
At node 5 ($L_1 = 0, L_2 = 0.5$): $\quad N_4 = 0.25b + 0.5e + f = 0.\quad$ (3.112)
At node 6 ($L_1 = 0.5, L_2 = 0$): $\quad N_4 = 0.25a + 0.5d + f = 0.\quad$ (3.113)
At node 3 ($L_1 = L_2 = 0$): $\quad N_4 = f = 0.\quad$ (3.114)

From Eqs. (3.109)–(3.114), we obtain

$$c = 4 \quad \text{and} \quad a = b = d = e = f = 0. \quad (3.115)$$

Thus, we get

$$N_4 = 4L_1L_2. \quad (3.116)$$

E. Shape Function N_5 Shape function N_5 is 1 at node 5 and is 0 at all other nodes. Since $L_1 = 0$ and $L_2 = L_3 = 0.5$ at node 5, N_5 should be a quadratic function of area coordinates L_2 and L_3. Thus,

$$N_5 = aL_2^2 + bL_3^2 + cL_2L_3 + dL_2 + eL_3 + f. \quad (3.117)$$

The following relations should hold:

At node 5 ($L_2 = L_3 = 0.5$): $\quad N_5 = 0.25a + 0.25b + 0.25c + 0.5d$
$\qquad\qquad\qquad\qquad\qquad\qquad + 0.5e + f = 1.\quad$ (3.118)
At node 2 ($L_2 = 1, L_3 = 0$): $\quad N_5 = a + d + f = 0.\quad$ (3.119)
At node 3 ($L_2 = 0, L_3 = 1$): $\quad N_5 = b + e + f = 0.\quad$ (3.120)
At node 4 ($L_2 = 0.5, L_3 = 0$): $\quad N_5 = 0.25a + 0.5d + f = 0.\quad$ (3.121)
At node 6 ($L_2 = 0, L_3 = 0.5$): $\quad N_5 = 0.25b + 0.5e + f = 0.\quad$ (3.122)
At node 1 ($L_2 = L_3 = 0$): $\quad N_5 = f = 0.\quad$ (3.123)

From Eqs. (3.118)–(3.123), we obtain

$$c = 4 \quad \text{and} \quad a = b = c = d = e = f = 0. \quad (3.124)$$

Thus, we get

$$N_5 = 4L_2L_3. \tag{3.125}$$

F. Shape Function N_6 The shape function N_6 is 1 at node 6 and is 0 at all other nodes. Since $L_1 = L_3 = 0.5$ and $L_2 = 0$ at node 6, N_6 should be a quadratic function of the area coordinates L_1 and L_3. Thus,

$$N_6 = aL_1^2 + bL_3^2 + cL_1L_3 + dL_1 + eL_3 + f. \tag{3.126}$$

The following relations should hold:

At node 6 ($L_1 = L_3 = 0.5$): $N_6 = 0.25a + 0.25b + 0.25c + 0.5d$
$$+ 0.5e + f = 1. \tag{3.127}$$
At node 1 ($L_1 = 1, L_3 = 0$): $N_6 = a + d + f = 0.$ (3.128)
At node 3 ($L_1 = 0, L_3 = 1$): $N_6 = b + e + f = 0.$ (3.129)
At node 4 ($L_1 = 0.5, L_3 = 0$): $N_6 = 0.25a + 0.5d + f = 0.$ (3.130)
At node 5 ($L_1 = 0, L_3 = 0.5$): $N_6 = 0.25b + 0.5e + f = 0.$ (3.131)
At node 2 ($L_1 = L_3 = 0$): $N_6 = f = 0.$ (3.132)

From Eqs. (3.127)–(3.132), we obtain

$$c = 4 \quad \text{and} \quad a = b = d = e = f = 0. \tag{3.133}$$

Thus, we get

$$N_6 = 4L_3L_1. \tag{3.134}$$

The above results can be summarized as

$$N_1 = L_1(2L_1 - 1), \tag{3.135}$$
$$N_2 = L_2(2L_2 - 1), \tag{3.136}$$
$$N_3 = L_3(2L_3 - 1), \tag{3.137}$$
$$N_4 = 4L_1L_2, \tag{3.138}$$
$$N_5 = 4L_2L_3, \tag{3.139}$$
$$N_6 = 4L_3L_1. \tag{3.140}$$

3.4 DERIVATION OF EIGENVALUE MATRIX EQUATIONS

Next, we move to the derivation of eigenvalue matrix equations for the E^x_{pq} and E^y_{pq} modes. As shown in Eqs. (2.75) and (2.76) in Marcatili's method, the wave equations are

E^x_{pq} mode:
$$\frac{\partial^2 E_x}{\partial x^2} + \frac{\partial^2 E_x}{\partial y^2} + (k_0^2 \varepsilon_r - \beta^2) E_x = 0, \quad (3.141)$$

E^y_{pq} mode:
$$\frac{\partial^2 H_x}{\partial x^2} + \frac{\partial^2 H_x}{\partial y^2} + (k_0^2 \varepsilon_r - \beta^2) H_x = 0. \quad (3.142)$$

Taking into consideration the continuity conditions at neighboring elements, which will be discussed later, we can rewrite the wave equation for the E^y_{pq} mode as

$$\frac{1}{\varepsilon_r}\left(\frac{\partial^2 H_x}{\partial x^2} + \frac{\partial^2 H_x}{\partial y^2}\right) + \left(k_0^2 - \frac{\beta^2}{\varepsilon_r}\right) H_x = 0. \quad (3.143)$$

Using Eqs. (3.141) and (3.142), we get the scalar wave equations [2, 3]

$$\eta^2\left(\frac{\partial^2 \phi}{\partial x^2} + \frac{\partial^2 \phi}{\partial y^2}\right) + (k_0^2 \xi^2 - \eta^2 \beta^2)\phi = 0, \quad (3.144)$$

where for the E^x_{pq} mode

$$\phi = E_x, \quad \eta^2 = 1, \quad \xi^2 = \varepsilon_r = n_r^2, \quad (3.145)$$

and for the E^y_{pq} mode

$$\phi = H_x, \quad \eta^2 = \frac{1}{\varepsilon_r} = \frac{1}{n_r^2}, \quad \xi^2 = 1. \quad (3.146)$$

To derive the eigenvalue matrix equation for these modes, we use the Galerkin method discussed in Section 3.2. After the analysis region is divided into many elements, the wave function ϕ_e at the nodes in e is

3.4 DERIVATION OF EIGENVALUE MATRIX EQUATIONS

expressed by shape functions N_1 and wave functions ϕ_{ei}. In other words, the wave function is expanded by the shape functions as

$$\phi_e = \sum_{i=1}^{M_e} N_{ei}\phi_{ei} = [N_e]^T\{\phi_e\}, \qquad (3.147)$$

where M_e is the number of nodes in e and T is a transposing operator for a matrix. We also used the following definitions:

$$[N_e] = [N_1 \ N_2 \ N_3 \cdots N_{M_e}]^T, \qquad (3.148)$$
$$\{\phi_e\} = (\phi_1 \ \phi_2 \ \phi_3 \cdots \phi_{M_e})^T, \qquad (3.149)$$

where the numbers of the nodes M_e in the first-order and second-order triangular elements are respectively 3 and 6.

Substituting Eq. (3.147) into the wave equation (3.144), we get

$$\eta_e^2\left(\frac{\partial^2}{\partial x^2} + \frac{\partial^2}{\partial y^2}\right)[N_e]^T\{\phi_e\} + (k_0^2\xi_e^2 - \eta_e^2\beta^2)[N_e]^T\{\phi_e\} = 0.$$

Multiplying the left-hand side of this equation by the shape function $[N_e]$ and integrating it in element e, we get

$$\iint_e [N_e]\eta_e^2\left(\frac{\partial^2}{\partial x^2} + \frac{\partial^2}{\partial y^2}\right)[N_e]^T \, dx \, dy\{\phi_e\}$$
$$+ \iint_e (k_0^2\xi_e^2 - \eta_e^2\beta^2)[N_e][N_e]^T \, dx \, dy\{\phi_e\} = \{0\}. \qquad (3.150)$$

Partially integrating the first term of Eq. (3.150) with respect to x and y, we get

$$\left[\int_{\Gamma_e} \eta_e^2[N_e]\frac{\partial[N_e]^T}{\partial x} \, dy + \int_{\Gamma_e} \eta_e^2[N_e]\frac{\partial[N_e]^T}{\partial y} \, dx\right]\{\phi_e\}$$
$$- \iint_e \eta_e^2\left\{\frac{\partial[N_e]}{\partial x}\frac{\partial[N_e]^T}{\partial x} + \frac{\partial[N_e]}{\partial y}\frac{\partial[N_e]^T}{\partial y}\right\} dx \, dy\{\phi_e\}$$
$$+ \iint_e (k_0^2\xi_e^2 - \eta_e^2\beta^2)[N_e][N_e]^T \, dx \, dy\{\phi_e\} = \{0\}.$$

86 FINITE-ELEMENT METHODS

Making use of Eq. (3.5), we get

$$\left[\int_{\Gamma_e} \eta_e^2 [N_e]^T \frac{\partial [N_e]^T}{\partial n} d\Gamma \right]\{\phi_e\}$$

$$- \iint_e \eta_e^2 \left\{ \frac{\partial [N_e]}{\partial x} \frac{\partial [N_e]^T}{\partial x} + \frac{\partial [N_e]}{\partial y} \frac{\partial [N_e]^T}{\partial y} \right\} dx\, dy\{\phi_e\}$$

$$+ \iint_e (k_0^2 \xi_e^2 - \eta_e^2 \beta^2)[N_e][N_e]^T\, dx\, dy\{\phi_e\} = \{0\}, \qquad (3.151)$$

where $\partial/\partial n$ is the derivative with respect to the outside normal and $\int_{\Gamma_e} d\Gamma$ is the line integration at boundary Γ_e. Since the method we are discussing is an FEM, it is necessary to sum up the contributions from all the elements:

$$\sum_e \left[\int_{\Gamma_e} \eta_e^2 [N_e] \frac{\partial [N_e]^T}{\partial n} d\Gamma \right]\{\phi_e\}$$

$$- \sum_e \iint_e \eta_e^2 \left\{ \frac{\partial [N_e]}{\partial x} \frac{\partial [N_e]^T}{\partial x} + \frac{\partial [N_e]}{\partial y} \frac{\partial [N_e]^T}{\partial y} \right\} dx\, dy\{\phi_e\}$$

$$+ \sum_e \iint_e (k_0^2 \xi_e^2 - \eta_e^2 \beta^2)[N_e][N_e]^T\, dx\, dy\{\phi_e\} = \{0\}. \qquad (3.152)$$

Here, we focus on the first term of the above equation. As mentioned before, $\int_\Gamma d\Gamma$ is a line integration at the boundary of element e. To simplify the argument here, we assume that the left-hand side of Eq. (3.152) is multiplied by the wave function vector $\{\phi_e\}^T$. Thus, we get the relation

$$\sum_e \left[\int_{\Gamma_e} \eta_e^2 [N_e] \frac{\partial [N_e]^T}{\partial n} d\Gamma \right]\{\phi_e\}$$

$$\rightarrow \sum_e \left[\int_{\Gamma_e} \eta_e^2 \underbrace{\{\phi_e\}^T [N_e]}_{\text{Wave function}} \underbrace{\frac{\partial [N_e]^T \{\phi_e\}}{\partial n}}_{\text{Derivative of wave function}} d\Gamma \right].$$

This means that the first term in Eq. (3.152) can be rewritten as

$$\sum_e \left[\int_{\Gamma_e} \eta_e^2 \phi_e \frac{\partial \phi_e}{\partial n} d\Gamma \right]$$
$$= \sum_e \left(\int_{\Gamma_{e+1}} \eta_{e+1}^2 \phi_{e+1} \frac{\partial \phi_{e+1}}{\partial n} d\Gamma - \int_{\Gamma_e} \eta_e^2 \phi_e \frac{\partial \phi_e}{\partial n} d\Gamma \right).$$

Here, we assume that the wave function ϕ_e and its normal derivative $\eta_e^2 \partial \phi_e / \partial n$ with constant η_e^2 are continuous at the boundaries with neighboring elements. Through this assumption, the line integral calculus terms inside the analysis region are canceled out, since, as shown in Fig. 3.9, the directions of the line integral calculus terms are opposite for each pair of neighboring elements. As a result, only the line integral calculus term of the periphery of the whole analysis region remains. Although this assumption is one of the limitations of the SC-FEM, it is a relatively good approximation. Thus, the first term in Eq. (3.152) can be rewritten as

$$\oint_\Gamma \eta^2 \phi \frac{\partial \phi}{\partial n} d\Gamma. \tag{3.153}$$

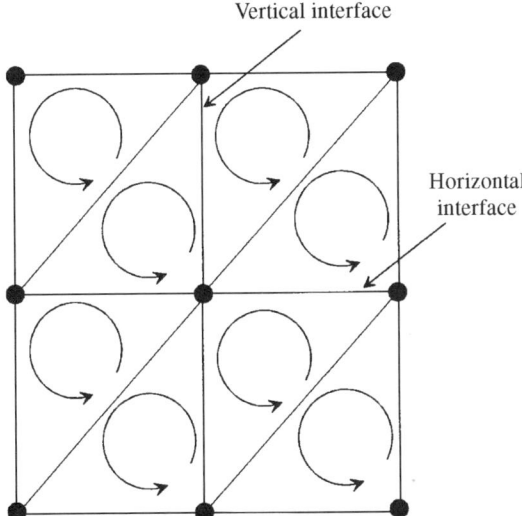

FIGURE 3.9. Canceling out of line integral calculus terms.

Substituting the line integral calculus term (3.153) into Eq. (3.152), we can reduce Eq. (3.152) to

$$([K] - \beta^2[M])\{\phi\} + \oint_\Gamma \eta^2 \phi \frac{\partial \phi}{\partial n} \, d\Gamma = \{0\}. \tag{3.154}$$

When we impose the Dirichlet or Neumann conditions on fields or their derivatives given by Eqs. (3.28) and (3.29), we can neglect the last term on the left-hand side of Eq. (3.154). Then Eq. (3.154) can be simplified to the eigenvalue matrix equation

$$([K] - \beta^2[M])\{\phi\} = \{0\}, \tag{3.155}$$

where the square of the propagation constant β is an eigenvalue and $\{\phi\}$ is an eigenvector. Here, we used the definitions

$$[K] = \sum_e \left\{ -\eta_e^2 \iint_e \left(\frac{\partial [N_e]}{\partial x} \frac{\partial [N_e]^T}{\partial x} + \frac{\partial [N_e]}{\partial y} \frac{\partial [N_e]^T}{\partial y} \right) dx \, dy \right.$$

$$\left. + k_0^2 \xi_e^2 \iint_e [N_e][N_e]^T \, dx \, dy \right\}, \tag{3.156}$$

$$[M] = \sum_e \eta_e^2 \iint_e [N_e][N_e]^T \, dx \, dy, \tag{3.157}$$

$$\{\phi\} = \sum_e \{\phi_e\}. \tag{3.158}$$

Since the relative permittivity ε_{re} is assumed to be constant in an element, η_e^2 and ξ_e^2 are also constant in the element.

The variable transformations

$$\bar{x} = xk_0, \tag{3.159}$$

$$\bar{y} = yk_0, \tag{3.160}$$

which are useful for suppressing the round-off errors in the calculation, enable the wave equation (3.155) to be reduced to

$$([\bar{K}] - n_{\text{eff}}^2[\bar{M}])\{\phi\} = \{0\}, \tag{3.161}$$

where

$$[\bar{K}] = \sum_e \left\{ -\eta_e^2 \int\int_e \left(\frac{\partial[N_e]}{\partial \bar{x}} \frac{\partial[N_e]^T}{\partial \bar{x}} + \frac{\partial[N_e]}{\partial \bar{y}} \frac{\partial[N_e]^T}{\partial \bar{y}} \right) d\bar{x}\, d\bar{y} \right.$$
$$\left. + \xi_e^2 \int\int_e [N_e][N_e]^T d\bar{x}\, d\bar{y} \right\}, \quad (3.162)$$

$$[\bar{M}] = \sum_e \eta_e^2 \int\int_e [N_e][N_e]^T d\bar{x}\, d\bar{y}, \quad (3.163)$$

$$\eta_{\text{eff}} = \frac{\beta}{k_0}. \quad (3.164)$$

3.5 MATRIX ELEMENTS

In this section, we discuss the matrix elements for the first- and second-order triangular elements. A way to form the global matrixes using matrix elements will be shown in the next section.

The eigenvalue equation given in Eqs. (3.155)–(3.158) are

$$([K] - \beta^2[M])\{\phi\} = \{0\}, \quad (3.165)$$

where

$$[K] = \sum_e \left\{ -\eta_e^2 \int\int_e \left(\frac{\partial[N_e]}{\partial x} \frac{\partial[N_e]^T}{\partial x} + \frac{\partial[N_e]}{\partial y} \frac{\partial[N_e]^T}{\partial y} \right) dx\, dy \right.$$
$$\left. + k_0^2 \xi_e^2 \int\int_e [N_e][N_e]^T dx\, dy \right\}, \quad (3.166)$$
$$= \sum_e \{-\eta_e^2([A_e] + [B_e]) + k_0^2 \xi_e^2 [C_e]\},$$

$$[M] = \sum_e \eta_e^2 \int\int_e [N_e][N_e]^T dx\, dy \quad (3.167)$$
$$= \sum_e \eta_e^2 [C_e],$$

$$\{\phi\} = \sum_e \{\phi_e\}. \quad (3.168)$$

As mentioned before, since the parameters η_e^2 and ξ_e^2 are constant in each element, only the following terms shown in Eqs. (3.165)–(3.167) have to be calculated to obtain explicit expressions for matrixes $[K]$ and $[M]$:

$$[A_e] = \iint_e \frac{\partial [N_e]}{\partial x} \frac{\partial [N_e]^T}{\partial x} \, dx \, dy, \qquad (3.169)$$

$$[B_e] = \iint_e \frac{\partial [N_e]}{\partial y} \frac{\partial [N_e]^T}{\partial y} \, dx \, dy, \qquad (3.170)$$

$$[C_e] = \iint_e [N_e][N_e]^T \, dx \, dy. \qquad (3.171)$$

Since the shape function $[N_e]$ is expressed by area coordinates L_1, L_2, and L_3, they are useful for performing the integral calculus in Eqs. (3.169)–(3.171). For later convenience, these coordinates are rewritten as

$$L_1 = \frac{Q_1(x - x_2) + R_1(y - y_2)}{2S_e}, \qquad (3.172)$$

$$L_2 = \frac{Q_2(x - x_3) + R_2(y - y_3)}{2S_e}, \qquad (3.173)$$

$$L_3 = \frac{Q_3(x - x_1) + R_3(y - y_1)}{2S_e}, \qquad (3.174)$$

where

$$Q_1 = y_2 - y_3, \qquad (3.175)$$
$$Q_2 = y_3 - y_1, \qquad (3.176)$$
$$Q_3 = y_1 - y_2, \qquad (3.177)$$
$$R_1 = x_3 - x_2, \qquad (3.178)$$
$$R_2 = x_1 - x_3, \qquad (3.179)$$
$$R_3 = x_2 - x_1, \qquad (3.180)$$

and

$$S_e = \tfrac{1}{2}[(y_3 - y_1)(x_2 - x_1) - (x_3 - x_1)(y_2 - y_1)]. \qquad (3.181)$$

Here, S_e is the area of the triangle 123 shown in Fig. 3.4. The derivatives of area coordinates L_1, L_2, and L_3 with respect to coordinates x and y are

$$\frac{\partial L_1}{\partial x} = \frac{Q_1}{2S_e}, \quad \frac{\partial L_2}{\partial x} = \frac{Q_2}{2S_e}, \quad \frac{\partial L_3}{\partial x} = \frac{Q_3}{2S_e}, \quad (3.182)$$

$$\frac{\partial L_1}{\partial y} = \frac{R_1}{2S_e}, \quad \frac{\partial L_2}{\partial y} = \frac{R_2}{2S_e}, \quad \frac{\partial L_3}{\partial y} = \frac{R_3}{2S_e}. \quad (3.183)$$

The following calculations will use the following convenient integration formula for the area coordinates:

$$\iint_e L_1^i L_2^j L_3^k \, dx \, dy = \frac{i! j! k!}{(i+j+k+2)!} 2S_e \quad (i, j, k = 0, 1, 2, 3, \ldots). \quad (3.184)$$

The derivation of this formula is shown in Appendix B.

3.5.1 First-Order Triangular Elements

Figure 3.4 shows a first-order triangular element. Since it has three nodes, its shape function $[N_e]$ has three components:

$$[N_e] = [N_1 \ N_2 \ N_3]^T. \quad (3.185)$$

As shown in Eq. (3.88), we have the following important relations between the shape functions N_1, N_2, and N_3 and the area coordinates L_1, L_2, and L_3:

$$N_1 = L_1, \quad N_2 = L_2, \quad N_3 = L_3. \quad (3.186)$$

The next step is to derive the explicit expressions of Eqs. (3.169) to (3.171).

A. $\iint_e (\partial [N_e]/\partial x)(\partial [N_e]^T/\partial x) \, dx \, dy$ The component representation of the matrix is

$$[A_e] = \iint_e \frac{\partial [N_e]}{\partial x} \frac{\partial [N_e]^T}{\partial x} \, dx \, dy = \begin{pmatrix} a_{11} & a_{12} & a_{13} \\ a_{21} & a_{22} & a_{23} \\ a_{31} & a_{32} & a_{33} \end{pmatrix}, \quad (3.187)$$

and the integrand is

$$\frac{\partial[N_e]}{\partial x}\frac{\partial[N_e]^T}{\partial x} = \begin{pmatrix} \partial N_1/\partial x \\ \partial N_2/\partial x \\ \partial N_3/\partial x \end{pmatrix} \begin{pmatrix} \dfrac{\partial N_1}{\partial x} & \dfrac{\partial N_2}{\partial x} & \dfrac{\partial N_3}{\partial x} \end{pmatrix}$$

$$= \begin{pmatrix} \left(\dfrac{\partial N_1}{\partial x}\right)^2 & \dfrac{\partial N_1}{\partial x}\dfrac{\partial N_2}{\partial x} & \dfrac{\partial N_1}{\partial x}\dfrac{\partial N_3}{\partial x} \\ \dfrac{\partial N_2}{\partial x}\dfrac{\partial N_1}{\partial x} & \left(\dfrac{\partial N_2}{\partial x}\right)^2 & \dfrac{\partial N_2}{\partial x}\dfrac{\partial N_3}{\partial x} \\ \dfrac{\partial N_3}{\partial x}\dfrac{\partial N_1}{\partial x} & \dfrac{\partial N_3}{\partial x}\dfrac{\partial N_2}{\partial x} & \left(\dfrac{\partial N_3}{\partial x}\right)^2 \end{pmatrix}. \quad (3.188)$$

Using the integration formula shown in Eq. (3.184), we obtain the following matrix elements:

$$a_{11} = \iint_e \left(\frac{\partial N_1}{\partial x}\right)^2 dx\,dy = \iint_e \left(\frac{\partial L_1}{\partial x}\right)^2 dx\,dy$$

$$= \left(\frac{Q_1}{2S_e}\right)^2 \iint_e dx\,dy = \frac{Q_1^2}{4S_e}, \quad (3.189)$$

$$a_{12} = \iint_e \left(\frac{\partial N_1}{\partial x}\right)\left(\frac{\partial N_2}{\partial x}\right) dx\,dy = \iint_e \left(\frac{\partial L_1}{\partial x}\right)\left(\frac{\partial L_2}{\partial x}\right) dx\,dy$$

$$= \left(\frac{Q_1}{2S_e}\right)\left(\frac{Q_2}{2S_e}\right) \iint_e dx\,dy = \frac{Q_1 Q_2}{4S_e} = a_{21}, \quad (3.190)$$

$$a_{13} = \iint_e \left(\frac{\partial N_1}{\partial x}\right)\left(\frac{\partial N_3}{\partial x}\right) dx\,dy = \iint_e \left(\frac{\partial L_1}{\partial x}\right)\left(\frac{\partial L_3}{\partial x}\right) dx\,dy$$

$$= \left(\frac{Q_1}{2S_e}\right)\left(\frac{Q_3}{2S_e}\right) \iint_e dx\,dy = \frac{Q_1 Q_3}{4S_e} = a_{31}, \quad (3.191)$$

$$a_{22} = \iint_e \left(\frac{\partial N_2}{\partial x}\right)^2 dx\,dy = \iint_e \left(\frac{\partial L_2}{\partial x}\right)^2 dx\,dy$$

$$= \left(\frac{Q_2}{2S_e}\right)^2 \iint_e dx\,dy = \frac{Q_2^2}{4S_e}, \quad (3.192)$$

3.5 MATRIX ELEMENTS

$$a_{23} = \iint_e \left(\frac{\partial N_2}{\partial x}\right)\left(\frac{\partial N_3}{\partial x}\right) dx\, dy = \iint_e \left(\frac{\partial L_2}{\partial x}\right)\left(\frac{\partial L_3}{\partial x}\right) dx\, dy$$

$$= \iint_e \left(\frac{Q_2}{2S_e}\right)\left(\frac{Q_3}{2S_e}\right) dx\, dy = \frac{Q_2 Q_3}{4S_e} = a_{32}, \quad (3.193)$$

$$a_{33} = \iint_e \left(\frac{\partial N_3}{\partial x}\right)^2 dx\, dy = \iint_e \left(\frac{\partial L_3}{\partial x}\right)^2 dx\, dy$$

$$= \iint_e \left(\frac{Q_3}{2S_e}\right)^2 dx\, dy = \frac{Q_3^2}{4S_e}. \quad (3.194)$$

B. $\iint_e (\partial[N_e]/\partial y)(\partial[N_e]^T/\partial y)\, dx\, dy$ The component representation of the matrix is

$$[B_e] = \iint_e \frac{\partial[N_e]}{\partial y} \frac{\partial[N_e]^T}{\partial y} dx\, dy = \begin{pmatrix} b_{11} & b_{12} & b_{13} \\ b_{21} & b_{22} & b_{23} \\ b_{31} & b_{32} & b_{33} \end{pmatrix}, \quad (3.195)$$

and the integrand is

$$\frac{\partial[N_e]}{\partial y} \frac{\partial[N_e]^T}{\partial y} = \begin{pmatrix} \left(\frac{\partial N_1}{\partial y}\right)^2 & \left(\frac{\partial N_1}{\partial y}\right)\left(\frac{\partial N_2}{\partial y}\right) & \left(\frac{\partial N_1}{\partial y}\right)\left(\frac{\partial N_3}{\partial y}\right) \\ \left(\frac{\partial N_2}{\partial y}\right)\left(\frac{\partial N_1}{\partial y}\right) & \left(\frac{\partial N_2}{\partial y}\right)^2 & \left(\frac{\partial N_2}{\partial y}\right)\left(\frac{\partial N_3}{\partial y}\right) \\ \left(\frac{\partial N_3}{\partial y}\right)\left(\frac{\partial N_1}{\partial y}\right) & \left(\frac{\partial N_3}{\partial y}\right)\left(\frac{\partial N_2}{\partial y}\right) & \left(\frac{\partial N_3}{\partial y}\right)^2 \end{pmatrix}.$$

$$(3.196)$$

Comparing Eq. (3.187) with Eq. (3.195), we find that the differences between them are the variables for the derivatives. Thus, the elements for the matrix in Eq. (3.195) can be obtained by simply substituting R_i for Q_i in Eqs. (3.189)–(3.194):

$$b_{11} = \iint_e \left(\frac{\partial N_1}{\partial y}\right)^2 dx\, dy = \iint_e \left(\frac{\partial L_1}{\partial y}\right)^2 dx\, dy = \frac{R_1^2}{4S_e}, \tag{3.197}$$

$$b_{12} = \iint_e \left(\frac{\partial N_1}{\partial y}\right)\left(\frac{\partial N_2}{\partial y}\right) dx\, dy = \iint_e \left(\frac{\partial L_1}{\partial y}\right)\left(\frac{\partial L_2}{\partial y}\right) dx\, dy$$

$$= \frac{R_1 R_2}{4S_e} = b_{21}, \tag{3.198}$$

$$b_{13} = \iint_e \left(\frac{\partial N_1}{\partial y}\right)\left(\frac{\partial N_3}{\partial y}\right) dx\, dy = \iint_e \left(\frac{\partial L_1}{\partial y}\right)\left(\frac{\partial L_3}{\partial y}\right) dx\, dy$$

$$= \frac{R_1 R_3}{4S_e} = b_{31}, \tag{3.199}$$

$$b_{22} = \iint_e \left(\frac{\partial N_2}{\partial y}\right)^2 dx\, dy = \iint_e \left(\frac{\partial L_2}{\partial y}\right)^2 dx\, dy = \frac{R_2^2}{4S_e}, \tag{3.200}$$

$$b_{23} = \iint_e \left(\frac{\partial N_2}{\partial y}\right)\left(\frac{\partial N_3}{\partial y}\right) dx\, dy = \iint_e \left(\frac{\partial L_2}{\partial y}\right)\left(\frac{\partial L_3}{\partial y}\right) dx\, dy$$

$$= \frac{R_2 R_3}{4S_e} = b_{32}, \tag{3.201}$$

$$b_{33} = \iint_e \left(\frac{\partial N_3}{\partial y}\right)^2 dx\, dy = \iint_e \left(\frac{\partial L_3}{\partial y}\right)^2 dx\, dy = \frac{R_3^2}{4S_e}. \tag{3.202}$$

C. $\iint_e [N_e][N_e]^T dx\, dy$ The component representation of the matrix is

$$[C_e] = \iint_e [N_e][N_e]^T dx\, dy = \begin{pmatrix} c_{11} & c_{12} & c_{13} \\ c_{21} & c_{22} & c_{23} \\ c_{31} & c_{32} & c_{33} \end{pmatrix}, \tag{3.203}$$

and the integrand is

$$[N_e][N_e]^T = \begin{pmatrix} N_1^2 & N_1 N_2 & N_1 N_3 \\ N_2 N_1 & N_2^2 & N_2 N_3 \\ N_3 N_1 & N_3 N_2 & N_3^2 \end{pmatrix}. \tag{3.204}$$

Using the integration formula shown in Eq. (3.184), we obtain the following matrix elements:

$$c_{11} = \iint_e N_1^2 \, dx \, dy = \iint_e L_1^2 \, dx \, dy = \frac{S_e}{6}, \tag{3.205}$$

$$c_{12} = \iint_e N_1 N_2 \, dx \, dy = \iint_e L_1 L_2 \, dx \, dy = \frac{S_e}{12} = c_{21}, \tag{3.206}$$

$$c_{13} = \iint_e N_1 N_3 \, dx \, dy = \iint_e L_1 L_3 \, dx \, dy = \frac{S_e}{12} = c_{31}, \tag{3.207}$$

$$c_{22} = \iint_e N_2^2 \, dx \, dy = \iint_e L_2^2 \, dx \, dy = \frac{S_e}{6}, \tag{3.208}$$

$$c_{23} = \iint_e N_2 N_3 \, dx \, dy = \iint_e L_2 L_3 \, dx \, dy = \frac{S_e}{12} = c_{32}, \tag{3.209}$$

$$c_{33} = \iint_e N_3^2 \, dx \, dy = \iint_e L_3^2 \, dx \, dy = \frac{S_e}{6}. \tag{3.210}$$

We have so far obtained matrixes $[A_e]$, $[B_e]$, and $[C_e]$ for the first-order triangular element e. We can form the global matrixes $[K]$ and $[M]$ by substituting matrixes $[A_e]$, $[B_e]$, and $[C_e]$ into Eqs. (3.166) and (3.167) and summing them up. Because the matrixes $[A_e]$, $[B_e]$, and $[C_e]$ are symmetrical 3×3 matrixes, the global matrixes are also symmetrical.

3.5.2 Second-Order Triangular Elements

Since the second-order triangular element (Fig. 3.8) has six nodes, its shape function $[N_e]$ has six components:

$$[N_e] = [N_1 \quad N_2 \quad N_3 \quad N_4 \quad N_5 \quad N_6]^T. \tag{3.211}$$

As shown in Eqs. (3.135)–(3.140), the important relations between shape functions $N_1, \ldots, N_6$ and area coordinates L_1, L_2, and L_3 are

$$N_1 = L_1(2L_1 - 1), \tag{3.212}$$
$$N_2 = L_2(2L_2 - 1), \tag{3.213}$$
$$N_3 = L_3(2L_3 - 1), \tag{3.214}$$
$$N_4 = 4L_1 L_2, \tag{3.215}$$
$$N_5 = 4L_2 L_3, \tag{3.216}$$
$$N_6 = 4L_3 L_1. \tag{3.217}$$

We derive the explicit expressions of Eqs. (3.169)–(3.171) for the second-order triangular element in a way similar to that in which we derived the corresponding expressions for the first-order triangular element.

A. $\iint_e (\partial [N_e]/\partial x)(\partial [N_e]^T/\partial x)\, dx\, dy$ The component representation of the matrix is

$$[A_e] = \iint_e \frac{\partial [N_e]}{\partial x} \frac{\partial [N_e]^T}{\partial x}\, dx\, dy = \begin{pmatrix} a_{11} & a_{12} & a_{13} & a_{14} & a_{15} & a_{16} \\ a_{21} & a_{22} & a_{23} & a_{24} & a_{25} & a_{26} \\ a_{31} & a_{32} & a_{33} & a_{34} & a_{35} & a_{36} \\ a_{41} & a_{42} & a_{43} & a_{44} & a_{45} & a_{46} \\ a_{51} & a_{52} & a_{53} & a_{54} & a_{55} & a_{56} \\ a_{61} & a_{62} & a_{63} & a_{64} & a_{65} & a_{66} \end{pmatrix},$$

(3.218)

and the integrand is

$$\frac{\partial [N_e]}{\partial x} \frac{\partial [N_e]^T}{\partial x}$$

$$= \begin{pmatrix} \frac{\partial N_1}{\partial x} & \frac{\partial N_2}{\partial x} & \frac{\partial N_3}{\partial x} & \frac{\partial N_4}{\partial x} & \frac{\partial N_5}{\partial x} & \frac{\partial N_6}{\partial x} \end{pmatrix}^T \begin{pmatrix} \frac{\partial N_1}{\partial x} & \frac{\partial N_2}{\partial x} & \frac{\partial N_3}{\partial x} & \frac{\partial N_4}{\partial x} & \frac{\partial N_5}{\partial x} & \frac{\partial N_6}{\partial x} \end{pmatrix}$$

$$= \begin{pmatrix} \left(\frac{\partial N_1}{\partial x}\right)^2 & \frac{\partial N_1}{\partial x}\frac{\partial N_2}{\partial x} & \frac{\partial N_1}{\partial x}\frac{\partial N_3}{\partial x} & \frac{\partial N_1}{\partial x}\frac{\partial N_4}{\partial x} & \frac{\partial N_1}{\partial x}\frac{\partial N_5}{\partial x} & \frac{\partial N_1}{\partial x}\frac{\partial N_6}{\partial x} \\ \frac{\partial N_2}{\partial x}\frac{\partial N_1}{\partial x} & \left(\frac{\partial N_2}{\partial x}\right)^2 & \frac{\partial N_2}{\partial x}\frac{\partial N_3}{\partial x} & \frac{\partial N_2}{\partial x}\frac{\partial N_4}{\partial x} & \frac{\partial N_2}{\partial x}\frac{\partial N_5}{\partial x} & \frac{\partial N_2}{\partial x}\frac{\partial N_6}{\partial x} \\ \frac{\partial N_3}{\partial x}\frac{\partial N_1}{\partial x} & \frac{\partial N_3}{\partial x}\frac{\partial N_2}{\partial x} & \left(\frac{\partial N_3}{\partial x}\right)^2 & \frac{\partial N_1}{\partial x}\frac{\partial N_4}{\partial x} & \frac{\partial N_1}{\partial x}\frac{\partial N_5}{\partial x} & \frac{\partial N_1}{\partial x}\frac{\partial N_6}{\partial x} \\ \frac{\partial N_4}{\partial x}\frac{\partial N_1}{\partial x} & \frac{\partial N_4}{\partial x}\frac{\partial N_2}{\partial x} & \frac{\partial N_4}{\partial x}\frac{\partial N_3}{\partial x} & \left(\frac{\partial N_4}{\partial x}\right)^2 & \frac{\partial N_4}{\partial x}\frac{\partial N_5}{\partial x} & \frac{\partial N_4}{\partial x}\frac{\partial N_6}{\partial x} \\ \frac{\partial N_5}{\partial x}\frac{\partial N_1}{\partial x} & \frac{\partial N_5}{\partial x}\frac{\partial N_2}{\partial x} & \frac{\partial N_5}{\partial x}\frac{\partial N_3}{\partial x} & \frac{\partial N_5}{\partial x}\frac{\partial N_4}{\partial x} & \left(\frac{\partial N_5}{\partial x}\right)^2 & \frac{\partial N_5}{\partial x}\frac{\partial N_6}{\partial x} \\ \frac{\partial N_6}{\partial x}\frac{\partial N_1}{\partial x} & \frac{\partial N_6}{\partial x}\frac{\partial N_2}{\partial x} & \frac{\partial N_6}{\partial x}\frac{\partial N_3}{\partial x} & \frac{\partial N_6}{\partial x}\frac{\partial N_4}{\partial x} & \frac{\partial N_6}{\partial x}\frac{\partial N_5}{\partial x} & \left(\frac{\partial N_6}{\partial x}\right)^2 \end{pmatrix}.$$

(3.219)

3.5 MATRIX ELEMENTS

The derivatives of the shape functions are

$$\frac{\partial N_i}{\partial x} = (4L_i - 1)\frac{\partial L_i}{\partial x} = \frac{Q_i}{2S_e}(4L_i - 1) \quad (i = 1, 2, 3), \tag{3.220}$$

$$\frac{\partial N_4}{\partial x} = 4\left(L_i\frac{\partial L_2}{\partial x} + L_2\frac{\partial L_i}{\partial x}\right) = \frac{2}{S_e}(Q_2L_1 + Q_1L_2), \tag{3.221}$$

$$\frac{\partial N_5}{\partial x} = 4\left(L_2\frac{\partial L_3}{\partial x} + L_3\frac{\partial L_2}{\partial x}\right) = \frac{2}{S_e}(Q_3L_2 + Q_2L_3), \tag{3.222}$$

$$\frac{\partial N_6}{\partial x} = 4\left(L_3\frac{\partial L_1}{\partial x} + L_1\frac{\partial L_3}{\partial x}\right) = \frac{2}{S_e}(Q_1L_3 + Q_3L_1). \tag{3.223}$$

Using the integration formula shown in Eq. (3.184) and the relations

$$\iint_e (4L_i - 1)L_j \, dx \, dy = \begin{cases} 0 & (i \neq j) \\ \frac{1}{3}S_e & (i = j) \end{cases} \tag{3.224}$$

$$\iint_e (4L_i - 1) \, dx \, dy = \frac{1}{3}S_e, \tag{3.225}$$

we obtain the following matrix elements for Eq. (3.218):

$$\begin{aligned}
a_{11} &= \iint_e \left(\frac{\partial N_1}{\partial x}\right)^2 dx\, dy = \left(\frac{Q_1}{2S_e}\right)^2 \iint_e (4L_1 - 1)^2 \, dx\, dy \\
&= \left(\frac{Q_1}{2S_e}\right)^2 \iint_e (16L_1^2 - 8L_1 + 1) \, dx\, dy \\
&= \left(\frac{Q_1}{2S_e}\right)^2 \left(16\frac{2!}{4!} - 8\frac{1}{3!} + \frac{1}{2!}\right) 2S_e \\
&= \left(\frac{Q_1}{2S_e}\right)^2 \left(16\frac{1}{4\cdot 3} - 8\frac{1}{3\cdot 2} + \frac{1}{2}\right) 2A_e \\
&= \left(\frac{Q_1}{2S_e}\right)^2 2A_e\left(\frac{4}{3} - \frac{4}{3} + \frac{1}{2}\right) = \frac{Q_1^2}{4S_e}, \tag{3.226}
\end{aligned}$$

$$a_{12} = \iint_e \left(\frac{\partial N_1}{\partial x}\right)\left(\frac{\partial N_2}{\partial x}\right) dx\, dy$$

$$= \left(\frac{Q_1}{2S_e}\right)\left(\frac{Q_2}{2S_e}\right) \iint_e (4L_1 - 1)(4L_2 - 1)\, dx\, dy$$

$$= \left(\frac{Q_1 Q_2}{4S_e^2}\right) \iint_e (16 L_1 L_2 - 4L_1 - 4L_2 + 1)\, dx\, dy$$

$$= \left(\frac{Q_1 Q_2}{4S_e^2}\right) 2S_e \left(16\frac{1}{4!} - 4\frac{1}{3!} - 4\frac{1}{3!} + \frac{1}{2!}\right)$$

$$= \left(\frac{Q_1 Q_2}{4S_e^2}\right) 2S_e \left(\frac{2}{3} - \frac{4}{3} + \frac{1}{2}\right) = -\frac{Q_1 Q_2}{12 S_e} = a_{21}, \qquad (3.227)$$

$$a_{13} = \iint_e \left(\frac{\partial N_1}{\partial x}\right)\left(\frac{\partial N_2}{\partial x}\right) dx\, dy$$

$$= \left(\frac{Q_1}{2S_e}\right)\left(\frac{Q_3}{2S_e}\right) \iint_e (4L_1 - 1)(4L_3 - 1)\, dx\, dy$$

$$= -\frac{Q_1 Q_3}{12 S_e} = a_{31}, \qquad (3.228)$$

$$a_{14} = \iint_e \left(\frac{\partial N_1}{\partial x}\right)\left(\frac{\partial N_4}{\partial x}\right) dx\, dy$$

$$= \left(\frac{Q_1}{2S_e}\right)\left(\frac{2}{S_e}\right) \iint_e (4L_1 - 1)(Q_2 L_1 + Q_1 L_2)\, dx\, dy$$

$$= \frac{Q_1 Q_2}{3 S_e} = a_{41}, \qquad (3.229)$$

$$a_{15} = \iint_e \left(\frac{\partial N_1}{\partial x}\right)\left(\frac{\partial N_5}{\partial x}\right) dx\, dy$$

$$= \left(\frac{Q_1}{2S_e}\right)\left(\frac{2}{S_e}\right) \iint_e (4L_1 - 1)(Q_3 L_2 + Q_2 L_3)\, dx\, dy = 0 = a_{51}, \qquad (3.230)$$

$$a_{16} = \iint_e \left(\frac{\partial N_1}{\partial x}\right)\left(\frac{\partial N_6}{\partial x}\right) dx\, dy$$

$$= \left(\frac{Q_1}{2S_e}\right)\left(\frac{2}{S_e}\right) \iint_e (4L_1 - 1)(Q_1 L_3 + Q_3 L_1)\, dx\, dy$$

$$= \frac{Q_1 Q_3}{3 S_e} = a_{61}, \qquad (3.231)$$

3.5 MATRIX ELEMENTS

$$a_{22} = \iint_e \left(\frac{\partial N_2}{\partial x}\right)^2 dx\,dy = \left(\frac{Q_2}{2A_e}\right)^2 \iint_e (4L_2 - 1)^2 \, dx\,dy = \frac{Q_2^2}{4S_e}, \quad (3.232)$$

$$a_{23} = \iint_e \left(\frac{\partial N_2}{\partial x}\right)\left(\frac{\partial N_3}{\partial x}\right) dx\,dy$$
$$= \left(\frac{Q_2}{2S_e}\right)\left(\frac{Q_3}{2S_e}\right) \iint_e (4L_2 - 1)(4L_3 - 1) \, dx\,dy$$
$$= -\frac{Q_2 Q_3}{12 S_e} = a_{32}, \quad (3.233)$$

$$a_{24} = \iint_e \left(\frac{\partial N_2}{\partial x}\right)\left(\frac{\partial N_4}{\partial x}\right) dx\,dy$$
$$= \left(\frac{Q_2}{2S_e}\right)\left(\frac{2}{S_e}\right) \iint_e (4L_2 - 1)(Q_2 L_1 + Q_1 L_2) \, dx\,dy$$
$$= \frac{Q_1 Q_2}{3 S_e} = a_{42}, \quad (3.234)$$

$$a_{25} = \iint_e \left(\frac{\partial N_2}{\partial x}\right)\left(\frac{\partial N_5}{\partial x}\right) dx\,dy$$
$$= \left(\frac{Q_2}{2S_e}\right)\left(\frac{2}{S_e}\right) \iint_e (4L_2 - 1)(Q_3 L_2 + Q_2 L_3) \, dx\,dy = \frac{Q_2 Q_3}{3 S_e}$$
$$= a_{52}, \quad (3.235)$$

$$a_{26} = \iint_e \left(\frac{\partial N_2}{\partial x}\right)\left(\frac{\partial N_6}{\partial x}\right) dx\,dy$$
$$= \left(\frac{Q_2}{2S_e}\right)\left(\frac{2}{S_e}\right) \iint_e (4L_2 - 1)(Q_1 L_3 + Q_3 L_1) \, dx\,dy = 0 = a_{62}, \quad (3.236)$$

$$a_{33} = \iint_e \left(\frac{\partial N_3}{\partial x}\right)^2 dx\,dy = \left(\frac{Q_3}{2S_e}\right)^2 \iint_e (4L_3 - 1)^2 \, dx\,dy = \frac{Q_3^2}{4S_e}, \quad (3.237)$$

$$a_{34} = \iint_e \left(\frac{\partial N_3}{\partial x}\right)\left(\frac{\partial N_4}{\partial x}\right) dx\,dy$$
$$= \left(\frac{Q_3}{2S_e}\right)\left(\frac{2}{S_e}\right) \iint_e (4L_3 - 1)(Q_2 L_1 + Q_1 L_2) \, dx\,dy = 0 = a_{43}, \quad (3.238)$$

$$a_{35} = \iint_e \left(\frac{\partial N_3}{\partial x}\right)\left(\frac{\partial N_5}{\partial x}\right) dx\,dy$$
$$= \left(\frac{Q_3}{2S_e}\right)\left(\frac{2}{S_e}\right) \iint_e (4L_3 - 1)(Q_3 L_2 + Q_2 L_3) \, dx\,dy$$
$$= \frac{Q_2 Q_3}{3 S_e} = a_{53}, \quad (3.239)$$

$$a_{36} = \iint_e \left(\frac{\partial N_3}{\partial x}\right)\left(\frac{\partial N_6}{\partial x}\right) dx\, dy$$

$$= \left(\frac{Q_3}{2S_e}\right)\left(\frac{2}{S_e}\right) \iint_e (4L_3 - 1)(Q_1 L_3 + Q_3 L_1)\, dx\, dy$$

$$= \frac{Q_3 Q_1}{3 S_e} = a_{63}, \tag{3.240}$$

$$a_{44} = \iint_e \left(\frac{\partial N_4}{\partial x}\right)^2 dx\, dy = \left(\frac{2}{A_e}\right)^2 \iint_e (Q_2 L_1 + Q_1 L_2)^2\, dx\, dy$$

$$= \left(\frac{2}{S_e}\right)^2 \iint_e (Q_2^2 L_1^2 + 2 Q_1 Q_2 L_1 L_2 + Q_1^2 L_2^2)\, dx\, dy$$

$$= \frac{2}{3 S_e}(Q_1^2 + Q_1 Q_2 + Q_2^2), \tag{3.241}$$

$$a_{45} = \iint_e \left(\frac{\partial N_4}{\partial x}\right)\left(\frac{\partial N_5}{\partial x}\right) dx\, dy$$

$$= \left(\frac{2}{A_e}\right)^2 \iint_e (Q_2 L_1 + Q_1 L_2)(Q_3 L_2 + Q_2 L_3)\, dx\, dy$$

$$= \frac{1}{3 S_e}(Q_2 Q_3 + Q_2^2 + 2 Q_1 Q_3 + Q_1 Q_2) = a_{54}, \tag{3.242}$$

$$a_{46} = \iint_e \left(\frac{\partial N_4}{\partial x}\right)\left(\frac{\partial N_6}{\partial x}\right) dx\, dy$$

$$= \left(\frac{2}{S_e}\right)^2 \iint_e (Q_2 L_1 + Q_1 L_2)(Q_1 L_3 + Q_3 L_1)\, dx\, dy$$

$$= \frac{1}{3 S_e}(Q_1 Q_2 + 2 Q_2 Q_3 + Q_1^2 + Q_1 Q_3) = a_{64}, \tag{3.243}$$

$$a_{55} = \iint_e \left(\frac{\partial N_5}{\partial x}\right)^2 dx\, dy = \left(\frac{2}{S_e}\right)^2 \iint_e (Q_3 L_2 + Q_2 L_3)^2\, dx\, dy$$

$$= \left(\frac{2}{S_e}\right)^2 \iint_e (Q_3^2 L_2^2 + 2 Q_2 Q_3 L_2 L_3 + Q_2^2 L_3^2)\, dx\, dy$$

$$= \frac{2}{3 S_e}(Q_3^2 + Q_2 Q_3 + Q_2^2), \tag{3.244}$$

$$a_{56} = \iint_e \left(\frac{\partial N_5}{\partial x}\right)\left(\frac{\partial N_6}{\partial x}\right) dx\, dy$$

$$= \left(\frac{2}{S_e}\right)^2 \iint_e (Q_3 L_2 + Q_2 L_3)(Q_1 L_3 + Q_3 L_1)\, dx\, dy$$

$$= \frac{1}{3S_e}(Q_1 Q_3 + Q_3^2 + 2 Q_1 Q_2 + Q_2 Q_3) = a_{65}, \qquad (3.245)$$

$$a_{66} = \iint_e \left(\frac{\partial N_6}{\partial x}\right)^2 dx\, dy = \left(\frac{2}{S_e}\right)^2 \iint_e (Q_1 L_3 + Q_3 L_1)^2\, dx\, dy$$

$$= \left(\frac{2}{S_e}\right)^2 \iint_e (Q_1^2 L_3 + 2 Q_1 Q_3 L_1 L_3 + Q_3^2 L_1^2)\, dx\, dy$$

$$= \frac{2}{3S_e}(Q_1^2 + Q_1 Q_3 + Q_3^2). \qquad (3.246)$$

B. $\iint_e (\partial [N_e]/\partial y)(\partial [N_e]^T/\partial y)\, dx\, dy$ The component representation of the matrix is

$$[B_e] = \iint_e \frac{\partial [N_e]}{\partial y} \frac{\partial [N_e]^T}{\partial y}\, dx\, dy = \begin{pmatrix} b_{11} & b_{12} & b_{13} & b_{14} & b_{15} & b_{16} \\ b_{21} & b_{22} & b_{23} & b_{24} & b_{25} & b_{26} \\ b_{31} & b_{32} & b_{33} & b_{34} & b_{35} & b_{36} \\ b_{41} & b_{42} & b_{43} & b_{44} & b_{45} & b_{46} \\ b_{51} & b_{52} & b_{53} & b_{54} & b_{55} & b_{56} \\ b_{61} & b_{62} & b_{63} & b_{64} & b_{65} & b_{66} \end{pmatrix}.$$

$$(3.247)$$

Comparing Eq. (3.219) with Eq. (3.247), we find that the differences between them are the variables for the derivatives. Thus, the elements for the matrix in Eq. (3.247) can be obtained by simply substituting R_i for Q_i in Eqs. (3.226)–(3.246):

$$b_{11} = \iint_e \left(\frac{\partial N_1}{\partial y}\right)^2 dx\, dy = \frac{R_1^2}{4 S_e}, \qquad (3.248)$$

$$b_{12} = \iint_e \left(\frac{\partial N_1}{\partial y}\right)\left(\frac{\partial N_2}{\partial y}\right) dx\, dy = -\frac{R_1 R_2}{12 S_e} = b_{21}, \qquad (3.249)$$

$$b_{13} = \iint_e \left(\frac{\partial N_1}{\partial y}\right)\left(\frac{\partial N_3}{\partial y}\right) dx\, dy = -\frac{R_1 R_3}{12 S_e} = b_{31}, \qquad (3.250)$$

$$b_{14} = \iint_e \left(\frac{\partial N_1}{\partial y}\right)\left(\frac{\partial N_4}{\partial y}\right) dx\, dy = \frac{R_1 R_2}{3 S_e} = b_{41}, \qquad (3.251)$$

$$b_{15} = \iint_e \left(\frac{\partial N_1}{\partial y}\right)\left(\frac{\partial N_5}{\partial y}\right) dx\, dy = 0 = b_{51}, \qquad (3.252)$$

$$b_{16} = \iint_e \left(\frac{\partial N_1}{\partial y}\right)\left(\frac{\partial N_6}{\partial y}\right) dx\, dy = \frac{R_1 R_3}{3 S_e} = b_{61}, \qquad (3.253)$$

$$b_{22} = \iint_e \left(\frac{\partial N_2}{\partial y}\right)^2 dx\, dy = \frac{R_2^2}{4 S_e}, \qquad (3.254)$$

$$b_{23} = \iint_e \left(\frac{\partial N_2}{\partial y}\right)\left(\frac{\partial N_3}{\partial y}\right) dx\, dy = -\frac{R_2 R_3}{12 S_e} = b_{32}, \qquad (3.255)$$

$$b_{24} = \iint_e \left(\frac{\partial N_2}{\partial y}\right)\left(\frac{\partial N_4}{\partial y}\right) dx\, dy = \frac{R_1 R_2}{3 S_e} = b_{42}, \qquad (3.256)$$

$$b_{25} = \iint_e \left(\frac{\partial N_2}{\partial y}\right)\left(\frac{\partial N_5}{\partial y}\right) dx\, dy = \frac{R_2 R_3}{3 S_e} = b_{52}, \qquad (3.257)$$

$$b_{26} = \iint_e \left(\frac{\partial N_2}{\partial y}\right)\left(\frac{\partial N_6}{\partial y}\right) dx\, dy = 0 = b_{62}, \qquad (3.258)$$

$$b_{33} = \iint_e \left(\frac{\partial N_3}{\partial y}\right)^2 dx\, dy = \frac{R_3^2}{4 S_e}, \qquad (3.259)$$

$$b_{34} = \iint_e \left(\frac{\partial N_3}{\partial y}\right)\left(\frac{\partial N_4}{\partial y}\right) dx\, dy = 0 = b_{43}, \qquad (3.260)$$

$$b_{35} = \iint_e \left(\frac{\partial N_3}{\partial y}\right)\left(\frac{\partial N_5}{\partial y}\right) dx\, dy = \frac{R_2 R_3}{3 S_e} = b_{53}, \qquad (3.261)$$

$$b_{36} = \iint_e \left(\frac{\partial N_3}{\partial y}\right)\left(\frac{\partial N_6}{\partial y}\right) dx\, dy = \frac{R_3 R_1}{3 S_e} = b_{63}, \qquad (3.262)$$

$$b_{44} = \iint_e \left(\frac{\partial N_4}{\partial y}\right)^2 dx\, dy = \frac{2}{3 S_e}(R_1^2 + R_1 R_2 + R_2^2), \qquad (3.263)$$

$$b_{45} = \iint_e \left(\frac{\partial N_4}{\partial y}\right)\left(\frac{\partial N_5}{\partial y}\right) dx\, dy$$

$$= \frac{1}{3 S_e}(R_2 R_3 + R_2^2 + 2 R_1 R_3 + R_1 R_2) = b_{54}, \qquad (3.264)$$

$$b_{46} = \iint_e \left(\frac{\partial N_6}{\partial y}\right)\left(\frac{\partial N_6}{\partial y}\right) dx\, dy$$
$$= \frac{1}{3S_e}(R_1R_2 + 2R_2R_3 + R_1^2 + R_1R_3) = b_{64}, \qquad (3.265)$$

$$b_{55} = \iint_e \left(\frac{\partial N_5}{\partial y}\right)^2 dx\, dy = \frac{2}{3S_e}(R_3^2 + R_2R_3 + R_2^2), \qquad (3.266)$$

$$b_{56} = \iint_e \left(\frac{\partial N_5}{\partial y}\right)\left(\frac{\partial N_6}{\partial y}\right) dx\, dy$$
$$= \frac{1}{3S_e}(R_1R_3 + R_3^2 + 2R_1R_2 + R_2R_3) = b_{65}, \qquad (3.267)$$

$$b_{66} = \iint_e \left(\frac{\partial N_6}{\partial y}\right)^2 dx\, dy = \frac{2}{3S_e}(R_1^2 + R_1R_3 + R_3^2). \qquad (3.268)$$

C. $\iint_e [N_e][N_e]^T dx\, dy$ The component representation of the matrix is

$$[C_e] = \iint_e [N_e][N_e]^T dx\, dy = \begin{pmatrix} c_{11} & c_{12} & c_{13} & c_{14} & c_{15} & c_{16} \\ c_{21} & c_{22} & c_{23} & c_{24} & c_{25} & c_{26} \\ c_{31} & c_{32} & c_{33} & c_{34} & c_{35} & c_{36} \\ c_{41} & c_{42} & c_{43} & c_{44} & c_{45} & c_{46} \\ c_{51} & c_{52} & c_{53} & c_{54} & c_{55} & c_{56} \\ c_{61} & c_{62} & c_{63} & c_{64} & c_{65} & c_{66} \end{pmatrix}, \qquad (3.269)$$

and the integrand is

$$[N_e][N_e]^T = \begin{pmatrix} N_1^2 & N_1N_2 & N_1N_3 & N_1N_4 & N_1N_5 & N_1N_6 \\ N_2N_1 & N_2^2 & N_2N_3 & N_2N_4 & N_2N_5 & N_2N_6 \\ N_3N_1 & N_3N_2 & N_3^2 & N_3N_4 & N_3N_5 & N_3N_6 \\ N_4N_1 & N_4N_2 & N_4N_3 & N_4^2 & N_4N_5 & N_4N_6 \\ N_5N_1 & N_5N_2 & N_5N_3 & N_5N_4 & N_5^2 & N_5N_6 \\ N_6N_1 & N_6N_2 & N_6N_3 & N_6N_4 & N_6N_5 & N_6^2 \end{pmatrix}. \qquad (3.270)$$

Using the integration formula shown in Eq. (3.184), we obtain the following matrix elements:

$$c_{11} = \iint_e N_1^2 dx\, dy = \iint_e L_1^2(2L_1 - 1)^2 dx\, dy = \frac{S_e}{30}, \qquad (3.271)$$

$$c_{12} = \iint_e N_1 N_2 dx\, dy = \iint_e (2L_1 - 1)L_1(2L_2 - 1)L_2\, dx\, dy$$
$$= -\frac{S_e}{180} = c_{21}, \qquad (3.272)$$

$$c_{13} = \iint_e N_1 N_3 \, dx \, dy = \iint_e (2L_1 - 1)L_1(2L_3 - 1)L_3 \, dx \, dy$$
$$= -\frac{S_e}{180} = c_{31}, \qquad (3.273)$$

$$c_{14} = \iint_e N_1 N_4 \, dx \, dy = \iint_e (2L_1 - 1)L_1 4 L_1 L_2 \, dx \, dy = 0 = c_{41}, \qquad (3.274)$$

$$c_{15} = \iint_e N_1 N_5 \, dx \, dy = \iint_e (2L_1 - 1)L_1 4 L_2 L_3 \, dx \, dy$$
$$= -\frac{S_e}{45} = c_{51}, \qquad (3.275)$$

$$c_{16} = \iint_e N_1 N_6 \, dx \, dy = \iint_e (2L_1 - 1)L_1 4 L_3 L_1 \, dx \, dy = 0 = c_{61}, \qquad (3.276)$$

$$c_{22} = \iint_e N_2^2 \, dx \, dy = \iint_e (2L_2 - 1)^2 L_2^2 \, dx \, dy = \frac{S_e}{30}, \qquad (3.277)$$

$$c_{23} = \iint_e N_2 N_3 \, dx \, dy = \iint_e (2L_2 - 1)L_2(2L_3 - 1)L_3 \, dx \, dy$$
$$= -\frac{S_e}{180} = c_{32}, \qquad (3.278)$$

$$c_{24} = \iint_e N_2 N_4 \, dx \, dy = \iint_e (2L_2 - 1)L_2 4 L_1 L_2 \, dx \, dy = 0 = c_{42}, \qquad (3.279)$$

$$c_{25} = \iint_e N_2 N_5 \, dx \, dy = \iint_e (2L_2 - 1)L_2 4 L_2 L_3 \, dx \, dy = 0 = c_{52}, \qquad (3.280)$$

$$c_{26} = \iint_e N_2 N_6 \, dx \, dy = \iint_e (2L_2 - 1)L_2 4 L_3 L_1 \, dx \, dy$$
$$= -\frac{S_e}{45} = c_{62}, \qquad (3.281)$$

$$c_{33} = \iint_e N_3^2 \, dx \, dy = \iint_e (2L_3 - 1)L_3^2 \, dx \, dy = \frac{S_e}{30}, \qquad (3.282)$$

$$c_{34} = \iint_e N_3 N_4 \, dx \, dy = \iint_e (2L_3 - 1)L_3 4 L_1 L_2 \, dx \, dy$$
$$= -\frac{S_e}{45} = c_{43}, \qquad (3.283)$$

$$c_{35} = \iint_e N_3 N_5 \, dx \, dy = \iint_e (2L_3 - 1)L_3 4 L_2 L_3 \, dx \, dy = 0 = c_{53}, \qquad (3.284)$$

$$c_{36} = \iint_e N_3 N_6 \, dx \, dy = \iint_e (2L_3 - 1)L_3 4 L_3 L_1 \, dx \, dy = 0 = c_{63}, \qquad (3.285)$$

$$c_{44} = \iint_e N_4^2 \, dx \, dy = \iint_e (4L_1L_2)^2 \, dx \, dy = \frac{8S_e}{45}, \tag{3.286}$$

$$c_{45} = \iint_e N_4 N_5 \, dx \, dy = \iint_e 4L_1L_2 4L_2L_3 \, dx \, dy = \frac{4S_e}{45} = c_{54}, \tag{3.287}$$

$$c_{46} = \iint_e N_4 N_6 \, dx \, dy = \iint_e 4L_1L_2 4L_3L_1 \, dx \, dy = \frac{4S_e}{45} = c_{64}, \tag{3.288}$$

$$c_{55} = \iint_e N_5^2 \, dx \, dy = \iint_e (4L_2L_3)^2 \, dx \, dy = \frac{8S_e}{45}, \tag{3.289}$$

$$c_{56} = \iint_e N_5 N_6 \, dx \, dy = \iint_e 4L_2L_3 4L_3L_1 \, dx \, dy = \frac{4S_e}{45} = c_{65}, \tag{3.290}$$

$$c_{66} = \iint_e N_6^2 \, dx \, dy = \iint_e (4L_3L_1)^2 \, dx \, dy = \frac{8S_e}{45}. \tag{3.291}$$

We have thus obtained the matrixes $[A_e]$, $[B_e]$, and $[C_e]$ for the second-order triangular element e. We can form the global matrixes $[K]$ and $[M]$ by substituting $[A_e]$, $[B_e]$, and $[C_e]$ into Eqs. (3.166) and (3.167) and summing them. Because $[A_e]$, $[B_e]$, and $[C_e]$ are symmetrical 6×6 matrixes, the global matrixes are also symmetrical.

3.6 PROGRAMMING

As described above, when using an FEM, we first obtain the matrix elements for element e. We then obtain the global matrixes for the eigenvalue matrix equation by summing the contributions of all the elements. This section discusses how computer programs based on the first- and second-order triangular elements can be written by using

$$([K] - \beta^2[M])\{\phi\} = \{0\}, \tag{3.292}$$

which is the eigenvalue matrix equation. Here, $[K]$, $[M]$, and $\{\phi\}$ are defined as

$$[K] = \sum_e \{-\eta_e^2([A_e] + [B_e]) + k_0^2 \xi_e^2 [C_e]\}, \tag{3.293}$$

$$[M] = \sum_e \eta_e^2 [C_e], \tag{3.294}$$

$$\{\phi\} = \sum_e \{\phi_e\} \tag{3.295}$$

and $[A_e]$, $[B_e]$, and $[C_e]$ as

$$[A_e] = \iint_e \frac{\partial [N_e]}{\partial x} \frac{\partial [N_e]^T}{\partial x} \, dx \, dy, \qquad (3.296)$$

$$[B_e] = \iint_e \frac{\partial [N_e]}{\partial y} \frac{\partial [N_e]^T}{\partial y} \, dx \, dy, \qquad (3.297)$$

$$[C_e] = \iint_e [N_e][N_e]^T \, dx \, dy. \qquad (3.298)$$

3.6.1 First-Order Triangular Elements

Figure 3.10 shows an example of an optical waveguide whose buried structure has been divided into 18 first-order triangular elements $e_1, \ldots, e_{18}$. The core with width W comprises two elements e_9 and e_{10}, and Fig. 3.11 shows the local coordinates for element e_9. In this figure, the local coordinates of the node numbers 6, 7, and 10—whose coordinates are (x_6, y_6), (x_7, xy_7), and (x_{10}, y_{10})—are respectively 1, 2, and 3. Thus, the node number can be determined from the element number and the local coordinate.

The actual programming flow is as follows:

1. Divide the whole analysis region into a number of meshes by using first-order triangular elements.

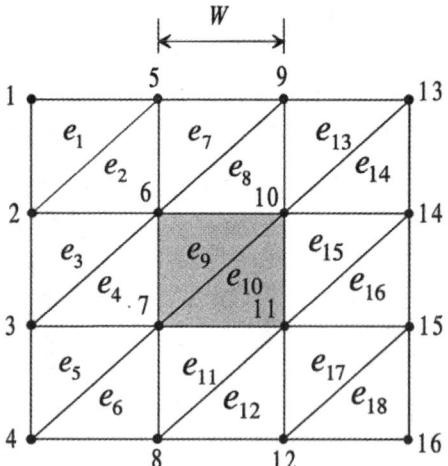

FIGURE 3.10. Mesh formed by first-order triangular elements.

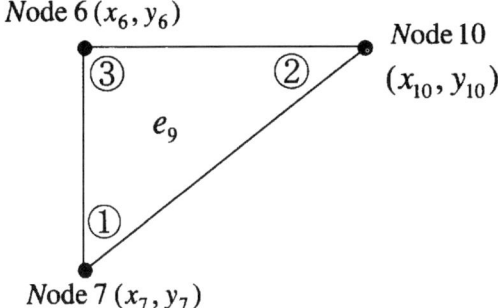

FIGURE 3.11. Local coordinates for the nodes of a first-order triangular element.

2. As shown in Fig. 3.11, any node number can be identified by specifying the element number and the local coordinate. For example, from the node number of element e_9 and the local coordinate 1, we get the node number 7 and the coordinate (x_7, y_7).
3. Using Eqs. (3.169)–(3.171), calculate the 3×3 matrixes $[A_e]$, $[B_e]$, and $[C_e]$ for each element e.
4. Add the calculated results for matrixes $[A_e]$, $[B_e]$, and $[C_e]$ to the global matrixes $[K]$ and $[M]$, whose row and column numbers correspond to the combinations of the element numbers and the local coordinates. For example, since the local coordinates 1 and 2 respectively correspond to nodes 7 and 10, the matrix elements with the first row and the second column of $[A_e]$, $[B_e]$, and $[C_e]$ are added to the matrix elements of the 7th row and the 10th column of both matrixes $[K]$ and $[M]$. Since the matrixes used in the scalar FEM are sparse and symmetrical, the amount of memory required for the matrixes can be reduced.
5. As shown in Fig. 3.10, node 6 belongs to six triangular elements (e_2, e_3, e_4, e_7, e_8, and e_9). Since each node belongs to more than one triangular element, obtain the global matrixes $[K]$ and $[M]$ by summing the calculated matrix elements of $[A_e]$, $[B_e]$, and $[C_e]$ for all triangular elements.
6. When imposing the boundary conditions, which will be discussed in Section 3.7, incorporate them into matrixes $[K]$ and $[M]$.
7. Obtain the propagation constant β or the effective index n_{eff} by solving the eigenvalue matrix equation (3.165).

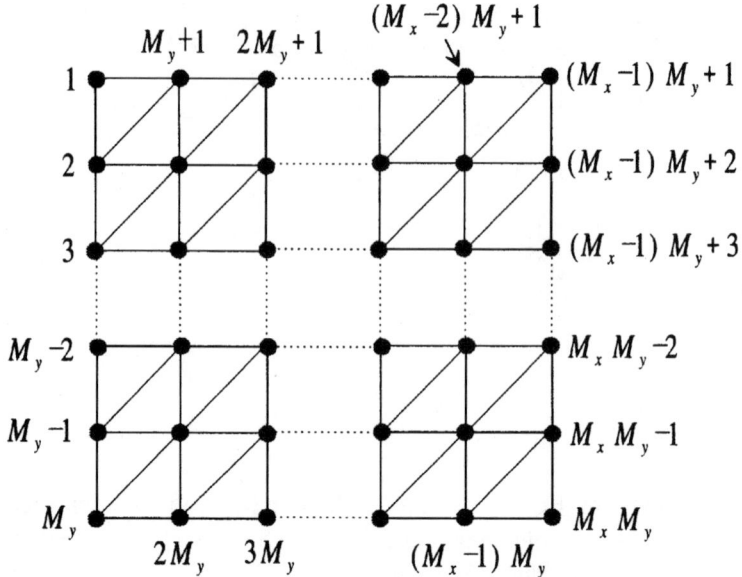

FIGURE 3.12. General meshes formed by first-order triangular elements.

Figure 3.12 shows an example of general meshes formed by first-order triangular elements. The numbers of nodes in the vertical and horizontal directions are respectively M_y and M_x, which means that the total number of nodes is $M_x \cdot M_y$. To discuss sparsity, we consider element e_1, in which there are three nodes; the numbers are 1, 2, and $M_y + 1$. Any node whose number is larger than $M_y + 1$ is not related to node 1.

Figure 3.13 shows matrixes $[K]$ and $[M]$ corresponding to the meshes shown in Fig. 3.12. The physical meanings of the matrix elements can be inferred from the wave functions $\phi_1, \phi_2, \ldots, \phi_M$ and in Fig. 3.13. Matrix element a_{ij} ($i \neq j$) is related to the interaction between wave functions ϕ_i and ϕ_j, and matrix element a_{ii} is related to the self-interaction of wave function ϕ_i. Thus, as shown in Fig. 3.13, matrixes $[K]$ and $[M]$ have nonzero elements until the $M_y + 1$ column. The number $M_y + 1$ is called the bandwidth of the sparse matrix.

3.6.2 Second-Order Triangular Elements

Figure 3.14 shows an example of general meshes formed by the second-order triangular elements, and Fig. 3.15 illustrates the correspondence

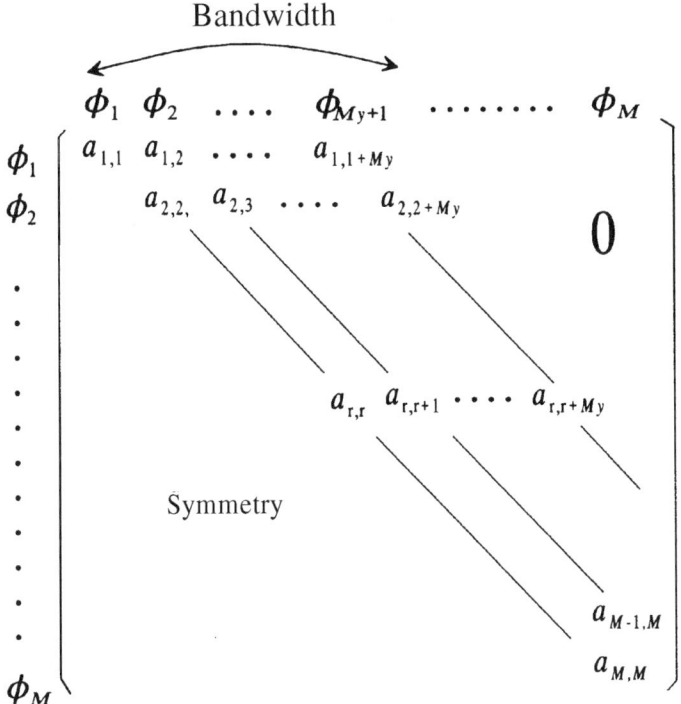

FIGURE 3.13. Forms of global matrixes $[K]$ and $[M]$, $M = M_x M_y$.

between the node numbers and the local coordinates for element e_1. A computer program based on second-order triangular elements is basically the same as one based on the first-order triangular elements, but it should be noted that each second-order triangular element has six local coordinates and that the corresponding matrixes $[K]$ and $[M]$ have a bandwidth of $2M_y + 1$.

When the number of the elements is the same, the total number of nodes and the bandwidth are larger for second-order triangular elements than they are for first-order triangular elements. Since the second-order elements approximate the unknown wave functions by quadratic curved surface functions, they are more accurate than the first-order elements, which approximate the unknown wave functions by plane functions. Thus, the second-order elements are numerically more efficient.

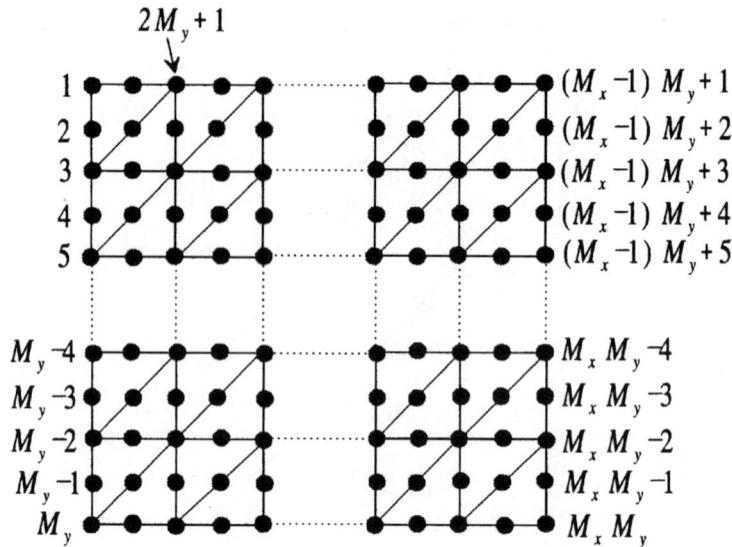

FIGURE 3.14. General meshes formed by second-order triangular elements.

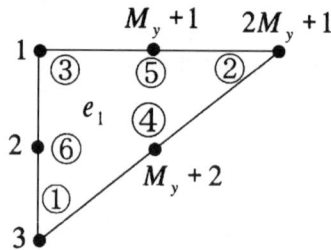

FIGURE 3.15. Local coordinates for the nodes of a second-order triangular element.

3.7 BOUNDARY CONDITIONS

In this section, we discuss the boundary conditions that should be applied to the nodes on the edges of the analysis region. The eigenvalue matrix equation was shown in Eq. (3.154) as

$$([K] - \beta^2[M])\{\phi\} + \oint_\Gamma \eta^2 \phi \frac{\partial \phi}{\partial n} \, d\Gamma = \{0\}. \tag{3.299}$$

Here, we discuss the two conditions used most widely.

3.7.1 Neumann Condition

The Neumann condition requires that the derivative of the wave function be set to zero, which means that the variation of the wave function at the boundaries would be negligibly small. Thus, we get

$$\frac{\partial \phi_i}{\partial n} = 0. \quad (3.300)$$

Substituting Eq. (3.300) into (3.299), we get the familiar eigenvalue matrix equation

$$([K] - \beta^2[M])\{\phi\} = \{0\}. \quad (3.301)$$

Here, we mention another important application of boundary conditions. Figure 3.10 shows the whole analysis region. Analyzing the whole region, we simultaneously obtain even modes including a dominant mode and odd modes whose fields are zero at the mirror-symmetrical plane of the structure. On the other hand, we can obtain the solutions for only the even modes or the odd modes by analyzing the half-plane structure (Fig. 3.16) with the Neumann condition or the Dirichlet condition applied at the mirror-symmetrical plane at the center. This is convenient when we analyze the definite modes of optical waveguides.

3.7.2 Dirichlet Condition

The Dirichlet condition requires that the wave functions at the boundaries be set to zero:

$$\phi_i = 0. \quad (3.302)$$

Thus, we also get the eigenvalue matrix equation (3.301). The Dirichlet condition requires a further process. Since Eq. (3.302) has to hold for the

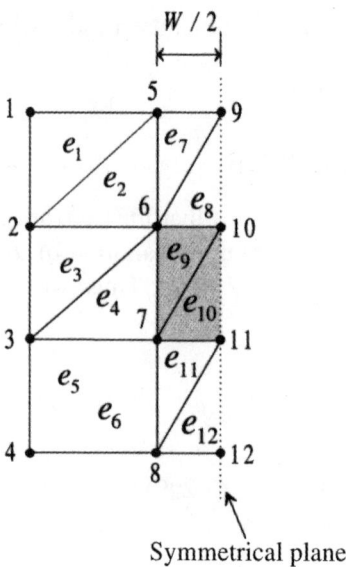

FIGURE 3.16. Boundary conditions on a mirror-symmetrical plane.

ith component ϕ_i of eigenvector $\{\phi\}$, some matrix elements other than the diagonal terms for matrixes $[K]$ and $[M]$ have to be set to zero:

$$\left(\begin{pmatrix} \ddots & & 0 & & \\ & \ddots & 0 & & \\ 0\cdots 0 & & K_{ii} & & 0\cdots 0 \\ & & 0 & \ddots & \\ & & 0 & & \ddots \end{pmatrix} - \beta^2 \begin{pmatrix} \ddots & & 0 & & \\ & \ddots & 0 & & \\ 0\cdots 0 & & M_{ii} & & 0\cdots 0 \\ & & 0 & \ddots & \\ & & 0 & & \ddots \end{pmatrix}\right) \begin{pmatrix} \phi_1 \\ \vdots \\ \phi_i \\ \vdots \\ \phi_M \end{pmatrix} = \{0\}. \tag{3.303}$$

Since the ith-row elements of the matrixes in Eq. (3.303) satisfy the equation

$$(K_{ii} - \beta^2 M_{ii})\phi_i = 0, \tag{3.304}$$

Eq. (3.302) holds.

Another way to implement the Dirichlet condition is to simply omit the nodes at the boundaries because under the Dirichlet condition they do not influence the other nodes. This requires less computer memory than is required when Eq. (3.303) is used.

PROBLEMS

1. In the derivation of the eigenvalue matrix equations (3.40) and (3.56), it was assumed that the wave function ϕ_e and its normal derivative $\eta_e^2 \, \partial \phi_e / \partial n$ are each continuous at the boundaries of two neighboring elements. This was the basis on which the line integral calculus terms in Eq. (3.51) were canceled out at the boundaries. Discuss the validity of the assumption.

ANSWER

a. E_{pq}^x **mode:**

$$\phi_e \rightarrow \begin{cases} \text{At horizontal interfaces } E_x \text{ is continuous.} \\ \text{At vertical interfaces } E_x \text{ is discontinuous.} \end{cases}$$

$$\eta_e^2 \frac{\partial \phi_e}{\partial n} \rightarrow \begin{cases} \text{At horizontal interfaces } \dfrac{\partial E_x}{\partial y} (\propto H_z) \text{ is continuous.} \\ \text{At vertical interfaces } \dfrac{\partial E_x}{\partial x} \left(\propto \dfrac{\varepsilon_{r2}}{\varepsilon_{r1}} E_z \right) \text{ is discontinuous.} \end{cases}$$

b. E_{pq}^y **mode:**

$$\phi_e \rightarrow \begin{cases} \text{At horizontal interfaces } H_x \text{ is continuous.} \\ \text{At vertical interfaces } H_x \text{ is discontinuous.} \end{cases}$$

$$\eta_e^2 \frac{\partial \phi_e}{\partial n} \rightarrow \begin{cases} \text{At horizontal interfaces } \dfrac{1}{\varepsilon_r} \dfrac{\partial H_x}{\partial y} (\propto E_z) \text{ is continuous.} \\ \text{At vertical interfaces } \dfrac{1}{\varepsilon_r} \dfrac{\partial H_x}{\partial x} \left(\propto \dfrac{1}{\varepsilon_r} H_z \right) \text{ is discontinuous.} \end{cases}$$

Thus, at the horizontal interfaces the assumption is valid for both the E_{pq}^x mode and the E_{pq}^x mode. At the vertical interfaces, on the other hand, the wave function ϕ_e and the derivative $\eta_e^2 \, \partial \phi_e / \partial n$ for the E_{pq}^x mode are discontinuous and the derivative $\eta_e^2 \, \partial \phi_e / \partial n$ for the E_{pq}^y mode is discontinuous.

From the above discussion, the conclusion is that *when the ratio of the width of the core to the thickness of the core is not large, which implies that the influences of the vertical boundaries cannot be neglected, the accuracy of the SC-FEM is degraded.* This is especially true for large-index-difference optical waveguides, such as those made from semiconductor materials.

114 FINITE-ELEMENT METHODS

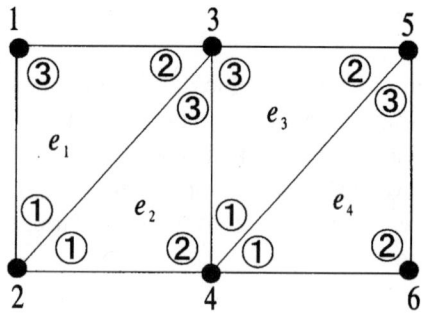

Node numbers: 1, 2, 3, ..., 6
Element numbers: e_1, e_2, e_3, e_4
Local coordinates: ①, ②, ③

FIGURE P3.1. Simple example of first-order triangular elements.

2. Figure P3.1 shows a simple example of first-order triangular elements, where the total node number is 6. Show the form of the matrix equation for this example.

ANSWER

$$\left(\begin{pmatrix} K_{11} & K_{12} & K_{13} & 0 & 0 & 0 \\ K_{21} & K_{22} & K_{23} & K_{24} & 0 & 0 \\ K_{31} & K_{32} & K_{33} & K_{34} & K_{35} & 0 \\ 0 & K_{42} & K_{43} & K_{44} & K_{45} & K_{46} \\ 0 & 0 & K_{53} & K_{54} & K_{55} & K_{56} \\ 0 & 0 & 0 & K_{64} & K_{65} & K_{66} \end{pmatrix}_{\text{Symmetry}} - \beta^2 \begin{pmatrix} M_{11} & M_{12} & M_{13} & 0 & 0 & 0 \\ M_{21} & M_{22} & M_{23} & M_{24} & 0 & 0 \\ M_{31} & M_{32} & M_{33} & M_{34} & M_{35} & 0 \\ 0 & M_{42} & M_{43} & M_{44} & M_{45} & M_{46} \\ 0 & 0 & M_{53} & M_{54} & M_{55} & M_{56} \\ 0 & 0 & 0 & M_{64} & M_{65} & M_{66} \end{pmatrix}_{\text{Symmetry}} \right) \begin{pmatrix} \phi_1 \\ \phi_2 \\ \phi_3 \\ \phi_4 \\ \phi_5 \\ \phi_6 \end{pmatrix} = \{0\}. \quad \text{(P3.1)}$$

REFERENCES

[1] O. C. Zienkiewitz, *The Finite Element Method*, 3rd ed., McGraw-Hill, New York, 1973.

[2] M. Koshiba, H. Saitoh, M. Eguchi, and K. Hirayama, "Simple scalar finite element approach to optical waveguides," *IEE Proc. J.*, vol. 139, pp. 166–171, 1992.

[3] M. Koshiba, *Optical Waveguide Theory by the Finite Element Method*, KTK Scientific Publishers and Kluwer Academic Publishers, Dordrecht, Holland, 1992.

[4] K. Kawano, S. Sekine, H. Takeuchi, M. Wada, M. Kohtoku, N. Yoshimoto, T. Ito, M. Yanagibashi, S. Kondo, and Y. Noguchi, "4×4 InGaAlAs/InAlAs MQW directional coupler waveguide switch modules integrated with spot-size converters and their 10 Gbit/s operation," *Electron. Lett.*, vol. 31, pp. 96–97, 1995.

[5] T. Itoh and R. Mittra, "Spectral domain approach for calculating the dispersion characteristics of microstrip lines," *IEEE Trans. Microwave Theory Tech.*, vol. MTT-21, pp. 496–499, 1973.

[6] K. Kawano, T. Kitoh, H. Jumonji, T. Nozawa, M. Yanagibashi, and T. Suzuki, "Spectral domain approach of coplanar waveguide traveling-wave electrodes and their applications to $Ti:LiNbO_3$ Mach–Zehnder optical modulators," *IEEE Trans. Microwave Theory Tech.*, vol. 39, pp. 1595–1601, 1991.

CHAPTER 4

FINITE-DIFFERENCE METHODS

Since the semivectorial finite-difference methods (SV-FDMs) developed by Stern [1, 2] are numerically efficient methods taking polarization into consideration and providing accurate results, they are widely used in the computer-aided design (CAD) software currently available for 2D cross-sectional analyses of optical waveguides. Finite-difference schemes are also used in the finite-difference beam propagation methods (FD-BPMs), which are of course also widely used in CAD software and which are discussed in detail in Chapter 5. The present chapter will help readers understand SV-FDMs and their formulation. It will also help readers become familiar with how to program them. Furthermore, it will teach users of CAD software not only how to identify the main causes of errors but also how to decrease the size of errors.

When the FEMs were discussed in Chapter 3, the wave equations themselves were not solved, but instead the functional was introduced and the variational principle was used, or the Galerkin method, which is a weighted residual method, was used. The FDMs dealt with in this chapter, in contrast, are more direct approaches to solving the wave equations. They solve eigenvalue matrix equations for electric fields or magnetic fields, equations derived from finite-difference approximations for the wave equations. This chapter briefly describes the finite-difference approximations and then derives the vectorial wave equations. It then obtains the semivectorial wave equations by ignoring the terms for the interaction between two polarized field components in the vectorial wave equations.

This chapter then discusses the formulation of the SV-FDMs, the errors caused by the finite-difference approximations, and SV-FDM programming. Although Stern's formulation uses equidistant discretization, this chapter uses the more versatile nonequidistant discretization.

4.1 FINITE-DIFFERENCE APPROXIMATIONS

In the FDMs discussed in this chapter, eigenvalue matrix equations are derived by using finite-difference schemes to approximate the wave equations. Let us first briefly examine the finite-difference approximations for the derivatives and then examine the accuracy of these approximations.

Assume that a 1D function $f(x)$ is continuous and smooth. As shown in Fig. 4.1, the function values f_1, f_2, and f_3 at $x = -h_1, h_2, 0$ are expressed as

$$f_1 = f(-h_1), \qquad (4.1)$$
$$f_2 = f(h_2), \qquad (4.2)$$
$$f_3 = f(0). \qquad (4.3)$$

Next, we can write f_1 and f_2 as Taylor series expansions around $x = 0$:

$$f_1 = f(-h_1) = f(0) - \frac{1}{1!}h_1 f^{(1)}(0) + \frac{1}{2!}h_1^2 f^{(2)}(0)$$
$$- \frac{1}{3!}h_1^3 f^{(3)}(0) + O(h_1^4), \qquad (4.4)$$
$$f_2 = f(h_2) = f(0) + \frac{1}{1!}h_2 f^{(1)}(0) + \frac{1}{2!}h_2^2 f^{(2)}(0)$$
$$+ \frac{1}{3!}h_2^3 f^{(3)}(0) + O(h_2^4), \qquad (4.5)$$

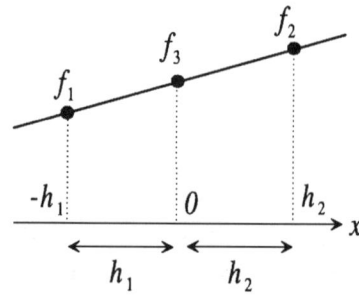

FIGURE 4.1. Difference approximations for derivatives.

where $f^{(n)}$ is an nth derivative defined as

$$f^{(n)}(x) = \frac{d^n f(x)}{dx^n}. \tag{4.6}$$

Subtracting Eq. (4.4) from Eq. (4.5), we get an expression for the first derivative:

$$f_2 - f_1 = (h_2 + h_1)f^{(1)}(0) + \frac{1}{2}(h_2^2 - h_1^2)f^{(2)}(0) + O(h^3).$$

Thus,

$$f^{(1)}(0) = \frac{f_2 - f_1}{h_1 + h_2} + \frac{1}{2}(h_2 - h_1)f^{(2)}(0) + O(h^2). \tag{4.7}$$

As shown in Eq. (4.7), the error caused by approximating the first derivative at $x = 0$ with the differential expression

$$f^{(1)}(0) = \frac{f_2 - f_1}{h_1 + h_2} \tag{4.8}$$

is $O(h^2)$ when $h_1 = h_2$ (equidistant discretization) and is $O(h)$ when $h_1 \neq h_2$ (nonequidistant discretization).

The error caused by approximating the second derivative with the expression

$$f^{(2)}(0) = \frac{2}{h_1 h_2} \frac{h_2 f_1 - (h_1 + h_2)f_3 + h_1 f_2}{(h_1 + h_2)} \tag{4.9}$$

is also $O(h^2)$ when $h_1 = h_2$ and is $O(h)$ when $h_1 \neq h_2$ (see Problem 1).

Thus, when CAD software is used, equidistant discretization is preferred whenever there is enough computer memory. When discretization is nonequidistant, we have to be careful that the ratio of h_1 to h_2 does not change drastically, since a drastic change would greatly increase the size of the errors.

4.2 WAVE EQUATIONS

4.2.1 Vectorial Wave Equations

In this section, the finite-difference expressions are obtained for the semivectorial wave equations. The approximations used in the process are made clear by starting from the fully vectorial forms of the equations.

The vectorial wave equation [Eq. (1.34)] for the electric field **E** is

$$\nabla^2 \mathbf{E} + \nabla\left(\frac{\nabla \varepsilon_r}{e_r} \cdot \mathbf{E}\right) + k_0^2 \varepsilon_r \mathbf{E} = \mathbf{0}. \tag{4.10}$$

Let us consider a structure uniform in the z direction. In this case, the derivative of relative permittivity with respect to z is zero:

$$\frac{\partial \varepsilon_r}{\partial z} = 0. \tag{4.11}$$

Thus, the second term in Eq. (4.10) can be written as

$$\nabla\left(\frac{\nabla \varepsilon_r}{\varepsilon_r} \cdot \mathbf{E}\right) = \nabla\left(\frac{1}{\varepsilon_r}\frac{\partial \varepsilon_r}{\partial x} E_x + \frac{1}{\varepsilon_r}\frac{\partial \varepsilon_r}{\partial y} E_y\right). \tag{4.12}$$

After substituting Eq. (4.12) into Eq. (4.10), we separate Eq. (4.10) into the x and y components. [The derivative with respect to z is indicated in Eqs. (4.13)–(4.30) in the form of $\partial/\partial z$ because that is the formulation used in the beam propagation methods dealt with in Chapter 5.] Thus, we obtain the vectorial wave equation using the electric field components E_x and E_y. Its x component is

$$\frac{\partial^2 E_x}{\partial x^2} + \frac{\partial}{\partial x}\left(\frac{1}{\varepsilon_r}\frac{\partial \varepsilon_r}{\partial x} E_x\right) + \frac{\partial^2 E_x}{\partial y^2} + \frac{\partial^2 E_x}{\partial z^2} + k_0^2 \varepsilon_r E_x + \frac{\partial}{\partial x}\left(\frac{1}{\varepsilon_r}\frac{\partial \varepsilon_r}{\partial y} E_y\right) = 0, \tag{4.13}$$

and its y component is

$$\frac{\partial^2 E_y}{\partial x^2} + \frac{\partial^2 E_y}{\partial y^2} + \frac{\partial}{\partial y}\left(\frac{1}{\varepsilon_r}\frac{\partial \varepsilon_r}{\partial y} E_y\right) + \frac{\partial^2 E_y}{\partial z^2} + k_0^2 \varepsilon_r E_y + \frac{\partial}{\partial y}\left(\frac{1}{\varepsilon_r}\frac{\partial \varepsilon_r}{\partial x} E_x\right) = 0. \tag{4.14}$$

Because

$$\frac{\partial}{\partial x}\left\{\frac{1}{\varepsilon_r}\frac{\partial}{\partial x}(\varepsilon_r E_x)\right\} = \frac{\partial^2 E_x}{\partial x^2} + \frac{\partial}{\partial x}\left(\frac{1}{\varepsilon_r}\frac{\partial \varepsilon_r}{\partial x}E_x\right) \quad (4.15)$$

and

$$\frac{\partial}{\partial y}\left\{\frac{1}{\varepsilon_r}\frac{\partial}{\partial y}(\varepsilon_r E_y)\right\} = \frac{\partial^2 E_y}{\partial y^2} + \frac{\partial}{\partial y}\left(\frac{1}{\varepsilon_r}\frac{\partial \varepsilon_r}{\partial y}E_y\right), \quad (4.16)$$

Eqs. (4.13) and (4.14) can be rewritten as

$$\frac{\partial}{\partial x}\left\{\frac{1}{\varepsilon_r}\frac{\partial}{\partial x}(\varepsilon_r E_x)\right\} + \frac{\partial^2 E_x}{\partial y^2} + \frac{\partial^2 E_x}{\partial z^2} + k_0^2 \varepsilon_r E_x + \frac{\partial}{\partial x}\left(\frac{1}{\varepsilon_r}\frac{\partial \varepsilon_r}{\partial y}E_y\right) = 0 \quad (4.17)$$

and

$$\frac{\partial^2 E_y}{\partial x^2} + \frac{\partial}{\partial y}\left\{\frac{1}{\varepsilon_r}\frac{\partial}{\partial y}(\varepsilon_r E_y)\right\} + \frac{\partial^2 E_y}{\partial z^2} + k_0^2 \varepsilon_r E_y + \frac{\partial}{\partial y}\left(\frac{1}{\varepsilon_r}\frac{\partial \varepsilon_r}{\partial x}E_x\right) = 0. \quad (4.18)$$

Let us next derive the corresponding components of the equation for the magnetic field **H**. The vectorial wave equation [Eq. (1.40)] for the magnetic field **H** is

$$\nabla^2 \mathbf{H} + \frac{\nabla \varepsilon_r}{\varepsilon_r} \times (\nabla \times \mathbf{H}) + k_0^2 \varepsilon_r \mathbf{H} = \mathbf{0}. \quad (4.19)$$

Here, the second term in Eq. (4.19) is investigated in detail. Recall that we are considering a structure uniform in the z direction and that Eq. (4.11) therefore holds. When **i**, **j**, and **k** are respectively assumed to be unit vectors in the x, y, and z directions, we can obtain

$$\nabla \varepsilon_r \times (\nabla \times \mathbf{H}) = \begin{vmatrix} \mathbf{i} & \mathbf{j} & \mathbf{k} \\ \dfrac{\partial \varepsilon_r}{\partial x} & \dfrac{\partial \varepsilon_r}{\partial y} & 0 \\ (\nabla \times \mathbf{H})_x & (\nabla \times \mathbf{H})_y & (\nabla \times \mathbf{H})_z \end{vmatrix}$$

$$= \frac{\partial \varepsilon_r}{\partial y}(\nabla \times \mathbf{H})_z \mathbf{i} - \frac{\partial \varepsilon_r}{\partial x}(\nabla \times \mathbf{H})_z \mathbf{j}$$

$$+ \left\{\frac{\partial \varepsilon_r}{\partial x}(\nabla \times \mathbf{H})_y - \frac{\partial \varepsilon_r}{\partial y}(\nabla \times \mathbf{H})_x\right\}\mathbf{k} \quad (4.20)$$

by assuming that $\partial \varepsilon_r / \partial z = 0$ and using the expressions

$$(\nabla \times \mathbf{H})_x = \frac{\partial H_z}{\partial y} - \frac{\partial H_y}{\partial z}, \qquad (4.21)$$

$$(\nabla \times \mathbf{H})_y = \frac{\partial H_x}{\partial z} - \frac{\partial H_z}{\partial x}, \qquad (4.22)$$

$$(\nabla \times \mathbf{H})_z = \frac{\partial H_y}{\partial x} - \frac{\partial H_x}{\partial y}. \qquad (4.23)$$

The substitution of Eqs. (4.21), (4.22), and (4.23) into Eq. (4.20) results in

$$\nabla \varepsilon_r \times (\nabla \times \mathbf{H}) = \frac{\partial \varepsilon_r}{\partial y}\left(\frac{\partial H_y}{\partial x} - \frac{\partial H_x}{\partial y}\right)\mathbf{i} - \frac{\partial \varepsilon_r}{\partial x}\left(\frac{\partial H_y}{\partial x} - \frac{\partial H_x}{\partial y}\right)\mathbf{j}$$
$$+ \left\{\frac{\partial \varepsilon_r}{\partial x}\left(\frac{\partial H_x}{\partial z} - \frac{\partial H_z}{\partial x}\right) - \frac{\partial \varepsilon_r}{\partial y}\left(\frac{\partial H_z}{\partial y} - \frac{\partial H_y}{\partial z}\right)\right\}\mathbf{k}. \quad (4.24)$$

Substituting Eq. (4.24) into Eq. (4.19) and separating the result into the x and y components, we obtain the vectorial wave equation using the magnetic field components H_x and H_y. Its x component is

$$\frac{\partial^2 H_x}{\partial x^2} + \frac{\partial^2 H_x}{\partial y^2} - \frac{1}{\varepsilon_r}\frac{\partial \varepsilon_r}{\partial y}\frac{\partial H_x}{\partial y} + \frac{\partial^2 H_x}{\partial z^2} + k_0^2 \varepsilon_r H_x + \frac{1}{\varepsilon_r}\frac{\partial \varepsilon_r}{\partial y}\frac{\partial H_y}{\partial x} = 0, \quad (4.25)$$

and its y component is

$$\frac{\partial^2 H_y}{\partial x^2} - \frac{1}{\varepsilon_r}\frac{\partial \varepsilon_r}{\partial x}\frac{\partial H_y}{\partial x} + \frac{\partial^2 H_y}{\partial y^2} + \frac{\partial^2 H_y}{\partial z^2} + k_0^2 \varepsilon_r H_y + \frac{1}{\varepsilon_r}\frac{\partial \varepsilon_r}{\partial x}\frac{\partial H_x}{\partial y} = 0. \quad (4.26)$$

And because

$$\varepsilon_r \frac{\partial}{\partial y}\left(\frac{1}{\varepsilon_r}\frac{\partial H_x}{\partial y}\right) = \frac{\partial^2 H_x}{\partial y^2} - \frac{1}{\varepsilon_r}\frac{\partial \varepsilon_r}{\partial y}\frac{\partial H_x}{\partial y} \qquad (4.27)$$

and

$$\varepsilon_r \frac{\partial}{\partial x}\left(\frac{1}{\varepsilon_r}\frac{\partial H_y}{\partial x}\right) = \frac{\partial^2 H_y}{\partial x^2} - \frac{1}{\varepsilon_r}\frac{\partial \varepsilon_r}{\partial x}\frac{\partial H_y}{\partial x}, \qquad (4.28)$$

Eqs. (4.25) and (4.26) can be rewritten as

$$\frac{\partial^2 H_x}{\partial x^2} + \varepsilon_r \frac{\partial}{\partial y}\left(\frac{1}{\varepsilon_r}\frac{\partial H_x}{\partial y}\right) + \frac{\partial^2 H_x}{\partial z^2} + k_0^2 \varepsilon_r H_x + \frac{1}{\varepsilon_r}\frac{\partial \varepsilon_r}{\partial y}\frac{\partial H_y}{\partial x} = 0 \quad (4.29)$$

and

$$\varepsilon_r \frac{\partial}{\partial x}\left(\frac{1}{\varepsilon_r}\frac{\partial H_y}{\partial x}\right) + \frac{\partial^2 H_y}{\partial y^2} + \frac{\partial^2 H_y}{\partial z^2} + k_0^2 \varepsilon_r H_y + \frac{1}{\varepsilon_r}\frac{\partial \varepsilon_r}{\partial x}\frac{\partial H_x}{\partial y} = 0. \quad (4.30)$$

Furthermore, since what we are concerned with here is a 2D cross-sectional analysis, the derivatives of the electric and magnetic fields with respect to z are constant:

$$\frac{\partial}{\partial z} = -j\beta, \quad (4.31)$$

where β is a propagation constant. Thus, using Eqs. (4.13) and (4.14), we obtain for the x component

$$\frac{\partial^2 E_x}{\partial x^2} + \frac{\partial}{\partial x}\left(\frac{1}{\varepsilon_r}\frac{\partial \varepsilon_r}{\partial x}E_x\right) + \frac{\partial^2 E_x}{\partial y^2} + (k_0^2 \varepsilon_r - \beta^2)E_x + \frac{\partial}{\partial x}\left(\frac{1}{\varepsilon_r}\frac{\partial \varepsilon_r}{\partial y}E_y\right) = 0 \quad (4.32)$$

and for the y component

$$\frac{\partial^2 E_y}{\partial x^2} + \frac{\partial^2 E_y}{\partial y^2} + \frac{\partial}{\partial y}\left(\frac{1}{\varepsilon_r}\frac{\partial \varepsilon_r}{\partial y}E_y\right) + (k_0^2 \varepsilon_r - \beta^2)E_y + \frac{\partial}{\partial y}\left(\frac{1}{\varepsilon_r}\frac{\partial \varepsilon_r}{\partial x}E_x\right) = 0. \quad (4.33)$$

They can be rewritten as

$$\frac{\partial}{\partial x}\left\{\frac{1}{\varepsilon_r}\frac{\partial}{\partial x}(\varepsilon_r E_x)\right\} + \frac{\partial^2 E_x}{\partial y^2} + (k_0^2 \varepsilon_r - \beta^2)E_x + \frac{\partial}{\partial x}\left(\frac{1}{\varepsilon_r}\frac{\partial \varepsilon_r}{\partial y}E_y\right) = 0 \quad (4.34)$$

and

$$\frac{\partial^2 E_y}{\partial x^2} + \frac{\partial}{\partial y}\left\{\frac{1}{\varepsilon_r}\frac{\partial}{\partial y}(\varepsilon_r E_y)\right\} + (k_0^2 \varepsilon_r - \beta^2)E_y + \frac{\partial}{\partial y}\left(\frac{1}{\varepsilon_r}\frac{\partial \varepsilon_r}{\partial x}E_x\right) = 0. \quad (4.35)$$

The last terms in Eqs. (4.32)–(4.35) correspond to the interactions between the x-directed electric field component E_x and the y-directed electric field component E_y.

Using Eqs. (4.25) and (4.26), on the other hand, we obtain for the x component

$$\frac{\partial^2 H_x}{\partial x^2} + \frac{\partial^2 H_x}{\partial y^2} - \frac{1}{\varepsilon_r}\frac{\partial \varepsilon_r}{\partial y}\frac{\partial H_x}{\partial y} + (k_0^2\varepsilon_r - \beta^2)H_x + \frac{1}{\varepsilon_r}\frac{\partial \varepsilon_r}{\partial y}\frac{\partial H_y}{\partial x} = 0 \quad (4.36)$$

and for the y component

$$\frac{\partial^2 H_y}{\partial x^2} - \frac{1}{\varepsilon_r}\frac{\partial \varepsilon_r}{\partial x}\frac{\partial H_y}{\partial x} + \frac{\partial^2 H_y}{\partial y^2} + (k_0^2\varepsilon_r - \beta^2)H_y + \frac{1}{\varepsilon_r}\frac{\partial \varepsilon_r}{\partial x}\frac{\partial H_x}{\partial y} = 0, \quad (4.37)$$

which can be rewritten as

$$\frac{\partial^2 H_x}{\partial x^2} + \varepsilon_r\frac{\partial}{\partial y}\left(\frac{1}{\varepsilon_r}\frac{\partial H_x}{\partial y}\right) + (k_0^2\varepsilon_r - \beta^2)H_x + \frac{1}{\varepsilon_r}\frac{\partial \varepsilon_r}{\partial y}\frac{\partial H_y}{\partial x} = 0 \quad (4.38)$$

and

$$\varepsilon_r\frac{\partial}{\partial x}\left(\frac{1}{\varepsilon_r}\frac{\partial H_y}{\partial x}\right) + \frac{\partial^2 H_y}{\partial y^2} + (k_0^2\varepsilon_r - \beta^2)H_y + \frac{1}{\varepsilon_r}\frac{\partial \varepsilon_r}{\partial x}\frac{\partial H_x}{\partial y} = 0. \quad (4.39)$$

Similar to what we saw in the corresponding equations for the electric field **E**, the last terms in Eqs. (4.36)–(4.39) correspond to the interactions between the x-directed magnetic field component H_x and the y-directed magnetic field component H_y.

4.2.2 Semivectorial Wave Equations

In equations for the fields that propagate in optical waveguides, the terms corresponding to the interaction between the x-directed electric field component E_x and the y-directed electric field component E_y,

$$\frac{\partial}{\partial x}\left(\frac{1}{\varepsilon_r}\frac{\partial \varepsilon_r}{\partial y}E_y\right) \quad \text{and} \quad \frac{\partial}{\partial y}\left(\frac{1}{\varepsilon_r}\frac{\partial \varepsilon_r}{\partial x}E_x\right) \quad \text{in Eqs. (4.32)–(4.35),}$$

and the terms corresponding to the interaction between the x-directed magnetic field component H_x and the y-directed magnetic field component H_y,

$$\frac{1}{\varepsilon_r}\frac{\partial \varepsilon_r}{\partial y}\frac{\partial H_y}{\partial x} \quad \text{and} \quad \frac{1}{\varepsilon_r}\frac{\partial \varepsilon_r}{\partial x}\frac{\partial H_x}{\partial y} \quad \text{in Eqs. (4.36)–(4.39),}$$

are usually small. Ignoring these terms for the interaction, we can decouple the vectorial wave equations for the x- and y-directed field components and reduce them to semivectorial wave equations, which can be solved in a way that is numerically efficient. Semivectorial analyses that neglect the terms for the interaction are therefore widely used when designing optical waveguide devices for which the coupling between the x- and y-directed polarizations does not have to be taken into consideration. As shown in Fig. 4.2, these analyses can be divided into the quasi-TE mode analysis, in which the principal field component is E_x or H_y, and the quasi-TM mode analysis, in which the principal field component is E_y or H_x.

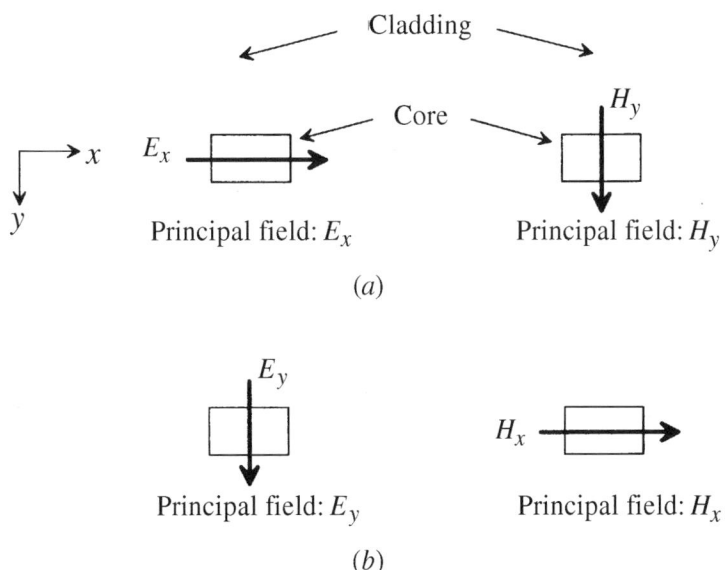

FIGURE 4.2. Principal field components for (*a*) quasi-TE and (*b*) quasi-TM modes.

A. Quasi-TE Mode The principal field component in the electric field representation for the quasi-TE mode, where the y-directed electric field component E_y is assumed to be zero, is the x-directed electric field component E_x. So according to Eq. (4.32), the semivectorial wave equation for the quasi-TE mode is

$$\frac{\partial^2 E_x}{\partial x^2} + \frac{\partial}{\partial x}\left(\frac{1}{\varepsilon_r}\frac{\partial \varepsilon_r}{\partial x}E_x\right) + \frac{\partial^2 E_x}{\partial y^2} + (k_0^2\varepsilon_r - \beta^2)E_x = 0, \tag{4.40}$$

which, according to Eq. (4.34), can be rewritten as

$$\frac{\partial}{\partial x}\left\{\frac{1}{\varepsilon_r}\frac{\partial}{\partial x}(\varepsilon_r E_x)\right\} + \frac{\partial^2 E_x}{\partial y^2} + (k_0^2\varepsilon_r - \beta^2)E_x = 0. \tag{4.41}$$

The principal field component in the magnetic field representation for the quasi-TE mode, on the other hand, where the x-directed magnetic field component H_x is assumed to be zero, is the y-directed magnetic field component H_y. So according to Eq. (4.37), the semivectorial wave equation is

$$\frac{\partial^2 H_y}{\partial x^2} - \frac{1}{\varepsilon_r}\frac{\partial \varepsilon_r}{\partial x}\frac{\partial H_y}{\partial x} + \frac{\partial^2 H_y}{\partial y^2} + (k_0^2\varepsilon_r - \beta^2)H_y = 0, \tag{4.42}$$

which, according to Eq. (4.39), can be rewritten as

$$\varepsilon_r \frac{\partial}{\partial x}\left(\frac{1}{\varepsilon_r}\frac{\partial H_y}{\partial x}\right) + \frac{\partial^2 H_y}{\partial y^2} + (k_0^2\varepsilon_r - \beta^2)H_y = 0. \tag{4.43}$$

B. Quasi-TM Mode The principal field component in the electric field representation for the quasi-TM mode, where the x-directed electric field component E_x is assumed to be zero, is the y-directed electric field component E_y. According to Eq. (4.33), the semivectorial wave equation for the quasi-TM mode is therefore

$$\frac{\partial^2 E_y}{\partial x^2} + \frac{\partial^2 E_y}{\partial y^2} + \frac{\partial}{\partial y}\left(\frac{1}{\varepsilon_r}\frac{\partial \varepsilon_r}{\partial y}E_y\right) + (k_0^2\varepsilon_r - \beta^2)E_y = 0, \tag{4.44}$$

which, according to Eq. (4.35), can be rewritten as

$$\frac{\partial^2 E_y}{\partial x^2} + \frac{\partial}{\partial y}\left\{\frac{1}{\varepsilon_r}\frac{\partial}{\partial y}(\varepsilon_r E_y)\right\} + (k_0^2\varepsilon_r - \beta^2)E_y = 0. \quad (4.45)$$

The principal field component for the quasi-TM mode in the magnetic field representation, where the y-directed magnetic field component H_y is assumed to be zero, is the x-directed magnetic field component H_x. As before, the semivectorial wave equation based on Eq. (4.36) is

$$\frac{\partial^2 H_x}{\partial x^2} + \frac{\partial^2 H_x}{\partial y^2} - \frac{1}{\varepsilon_r}\frac{\partial \varepsilon_r}{\partial y}\frac{\partial H_x}{\partial y} + (k_0^2\varepsilon_r - \beta^2)H_x = 0, \quad (4.46)$$

which, according to Eq. (4.38), can be rewritten as

$$\frac{\partial^2 H_x}{\partial x^2} + \varepsilon_r \frac{\partial}{\partial y}\left(\frac{1}{\varepsilon_r}\frac{\partial H_x}{\partial y}\right) + (k_0^2\varepsilon_r - \beta^2)H_x = 0. \quad (4.47)$$

4.2.3 Scalar Wave Equation

In the vectorial and semivectorial wave equations discussed above, the derivatives of relative permittivity ε_r with respect to the x and y coordinates are taken into consideration. If we assume these derivatives to be zero, or that

$$\frac{\partial \varepsilon_r}{\partial x} = 0 \quad \text{and} \quad \frac{\partial \varepsilon_r}{\partial y} = 0, \quad (4.48)$$

the wave equations can be reduced to the scalar wave equation

$$\frac{\partial^2 \phi}{\partial x^2} + \frac{\partial^2 \phi}{\partial y^2} + (k_0^2\varepsilon_r - \beta^2)\phi = 0, \quad (4.49)$$

where ϕ is a wave function and designates a scalar field.

4.3 FINITE-DIFFERENCE EXPRESSIONS OF WAVE EQUATIONS

We can derive the finite-difference expressions for the semivectorial wave equations by using the finite differences discussed in Section 4.1 to approximate the derivatives of the semivectorial wave equations for the electric field or the magnetic field representations for the quasi-TE and quasi-TM modes. The currently available CAD software for 2D cross-sectional analyses and 3D beam propagation analyses makes use of the finite-difference expressions discussed in this section.

Figure 4.3 shows a discretization used in a 2D cross-sectional analysis of optical waveguides. Here, the pair (p, q) is assumed to correspond to the (x, y) coordinates of a node. It should be noted that the interface of two materials is set midway between two nodes in order to minimize the error caused by the difference approximation. Although the discretization used by Stern [1, 2] was an equidistant one, the more versatile scheme described here is nonequidistant, with discretization widths e and w in the x direction and n and s in the y direction. The scalar case is also briefly discussed here. Although the variable transformations $\bar{x} = xk_0$ and $\bar{y} = yk_0$ are useful for reducing the round-off error in an actual calculation, they are not used here for simplicity. Reasonable results for the facet reflectivities of 3D semiconductor optical waveguides [3] have been obtained by incorporating the semivectorial finite-difference expressions discussed in this section into the bidirectional method of line BPM (MoL-BPM) [4].

4.3.1 Quasi-TE Mode

A. E_x Representation The E_x representation wave equation derived for the quasi-TE mode [Eq. (4.40)] is

$$\frac{\partial^2 E_x}{\partial x^2} + \frac{\partial}{\partial x}\left(\frac{1}{\varepsilon_r}\frac{\partial \varepsilon_r}{\partial x} E_x\right) + \frac{\partial^2 E_x}{\partial y^2} + (k_0^2 \varepsilon_r - \beta^2) E_x = 0. \tag{4.50}$$

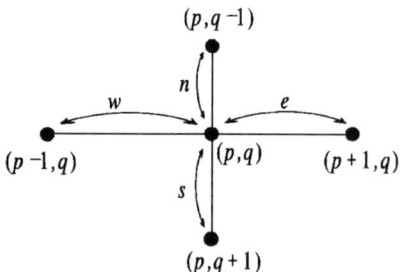

FIGURE 4.3. Nonequidistant discretization for the finite-difference method.

4.3 FINITE-DIFFERENCE EXPRESSIONS OF WAVE EQUATIONS

What we want to do now is derive the finite-difference expression for this equation. As shown in Fig. 4.3, the field components and coordinates at nodes are expressed as

$$E_{p+1,q} = E(x_{p+1}, y_q), \tag{4.51}$$

$$E_{p,q} = E(x_p, y_q), \tag{4.52}$$

$$E_{p-1,q} = E(x_{p-1}, y_q), \tag{4.53}$$

$$E_{p,q+1} = E(x_p, y_{q+1}), \tag{4.54}$$

$$E_{p,q-1} = E(x_p, y_{q-1}), \tag{4.55}$$

$$n = y_q - y_{q-1}, \tag{4.56}$$

$$s = y_{q+1} - y_q, \tag{4.57}$$

$$e = x_{p+1} - x_p, \tag{4.58}$$

$$w = x_p - x_{p-1}. \tag{4.59}$$

Using for $E_{p+1,q}$, $E_{p-1,q}$, $E_{p,q+1}$, and $E_{p,q-1}$ Taylor series expansions around (p, q), we get

$$E_{p+1,q} = E_{p,q} + \frac{1}{1!}\left.\frac{\partial E}{\partial x}\right|_{p,q} \cdot e + \frac{1}{2!}\left.\frac{\partial^2 E}{\partial x^2}\right|_{p,q} \cdot e^2 + O(e^3), \tag{4.60}$$

$$E_{p-1,q} = E_{p,q} - \frac{1}{1!}\left.\frac{\partial E}{\partial x}\right|_{p,q} \cdot w + \frac{1}{2!}\left.\frac{\partial^2 E}{\partial x^2}\right|_{p,q} \cdot w^2 + O(w^3), \tag{4.61}$$

$$E_{p,q+1} = E_{p,q} + \frac{1}{1!}\left.\frac{\partial E}{\partial y}\right|_{p,q} \cdot s + \frac{1}{2!}\left.\frac{\partial^2 E}{\partial y^2}\right|_{p,q} \cdot s^2 + O(s^3), \tag{4.62}$$

$$E_{p,q-1} = E_{p,q} - \frac{1}{1!}\left.\frac{\partial E}{\partial y}\right|_{p,q} \cdot n + \frac{1}{2!}\left.\frac{\partial^2 E}{\partial y^2}\right|_{p,q} \cdot n^2 + O(n^3). \tag{4.63}$$

First, we derive the finite-difference expression for the first term in Eq. (4.50). Multiplying Eq. (4.60) by w and Eq. (4.61) by e and then adding them, we find that

$$wE_{p+1,q} + eE_{p-1,q} = (w+e)E_{p,q} + \left.\frac{\partial^2 E}{\partial x^2}\right|_{p,q} \frac{e^2 w + w^2 e}{2!}$$

$$+ \left.\frac{\partial^3 E}{\partial x^3}\right|_{p,q} \frac{e^3 w - w^3 e}{3!} + \left.\frac{\partial^4 E}{\partial x^4}\right|_{p,q} \frac{e^4 w + w^4 e}{4!} + \cdots.$$

Therefore

$$\left.\frac{\partial^2 E}{\partial x^2}\right|_{p,q} = \frac{2!}{ew(e+w)}\{wE_{p+1,q} + eE_{p-1,q} - (e+w)E_{p,q}\}$$

$$- \left.\frac{\partial^3 E}{\partial x^3}\right|_{p,q} \frac{2!}{ew(e+w)} \frac{ew(e+w)(e-w)}{3!}$$

$$- \left.\frac{\partial^4 E}{\partial x^4}\right|_{p,q} \frac{2!}{ew(e+w)} \frac{ew(e+w)(e^2 - ew + w^2)}{4!} + \cdots$$

$$= \frac{2}{ew(e+w)}\{wE_{p+1,q} + eE_{p-1,q} - (e+w)E_{p,q}\}$$

$$- \left.\frac{\partial^3 E}{\partial x^3}\right|_{p,q} \frac{e-w}{3} - \left.\frac{\partial^4 E}{\partial x^4}\right|_{p,q} \frac{e^2 - ew + w^2}{12} + \cdots \quad (4.64)$$

and

$$\left.\frac{\partial^2 E}{\partial x^2}\right|_{p,q} = \frac{2}{e(e+w)}E_{p+1,q} + \frac{2}{w(e+w)}E_{p-1,q} - \frac{2}{ew}E_{p,q}. \quad (4.65)$$

As shown in Section 4.1, the error caused by the finite-difference approximation in Eq. (4.65) is $O(e-w)$ when $e \neq w$ (nonequidistant discretization) and is $O((\Delta x)^2)$ when $e = w = \Delta x$ (equidistant discretization).

We use a similar procedure to obtain the finite-difference expression for the derivative with respect to y [the third term in Eq. (4.50)]. Multiplying Eq. (4.62) by n and Eq. (4.63) by s and adding them, we get

$$nE_{p,q+1} + sE_{p,q-1} = (n+s)E_{p,q} + \left.\frac{\partial^2 E}{\partial y^2}\right|_{p,q} \frac{s^2 n + n^2 s}{2!}$$

$$+ \left.\frac{\partial^3 E}{\partial y^3}\right|_{p,q} \frac{s^3 n - n^3 s}{3!} + \left.\frac{\partial^4 E}{\partial y^4}\right|_{p,q} \frac{s^4 n + n^4 s}{4!} + \cdots.$$

Therefore

$$\left.\frac{\partial^2 E}{\partial y^2}\right|_{p,q} = \frac{2!}{ns(n+s)}\{nE_{p,q+1} + sE_{p,q-1} - (n+s)E_{p,q}\}$$

$$- \left.\frac{\partial^3 E}{\partial y^3}\right|_{p,q} \frac{2!}{ns(n+s)} \frac{ns(s-n)(s+n)}{3!}$$

$$- \left.\frac{\partial^4 E}{\partial y^4}\right|_{p,q} \frac{2!}{ns(n+s)} \frac{ns(n+s)(n^2 - ns + s^2)}{4!} + \cdots \quad (4.66)$$

and

$$\frac{\partial^2 E}{\partial y^2}\bigg|_{p,q} = \frac{2}{s(n+s)}E_{p,q+1} + \frac{2}{n(n+s)}E_{p,q-1} - \frac{2}{sn}E_{p,q}. \quad (4.67)$$

According to Eq. (4.66), the error caused by the finite-difference approximation in Eq. (4.67) is $O(s-n)$ when $n \neq s$ (nonequidistant discretization) and is $O((\Delta y)^2)$ when $s = n = \Delta y$ (equidistant discretization).

Now, let us consider the second term in Eq. (4.50), which includes a derivative of relative permittivity ε_r. For simplicity here, expression (4.6) is used to represent derivatives such as $\partial/\partial x$ and $\partial^2/\partial x^2$.

The difference center of the equations under discussion is (p,q). Here, we introduce $(p+\frac{1}{2}, q)$ as a hypothetical difference center between nodes (p,q) and $(p+1,q)$ and introduce $(p-\frac{1}{2}, q)$ as a hypothetical difference center between nodes $(p-1,q)$ and (p,q). Using these two hypothetical difference centers, we can get the following two Taylor series expansions:

$$\left(\frac{1}{\varepsilon_r}\frac{\partial \varepsilon_r}{\partial x}E_x\right)_{p+1/2,q} = \left(\frac{1}{\varepsilon_r}\frac{\partial \varepsilon_r}{\partial x}E_x\right)_{p,q} + \frac{1}{1!}\left(\frac{1}{\varepsilon_r}\frac{\partial \varepsilon_r}{\partial x}E_x\right)^{(1)}_{p,q}\left(\frac{e}{2}\right)$$

$$+ \frac{1}{2!}\left(\frac{1}{\varepsilon_r}\frac{\partial \varepsilon_r}{\partial x}E_x\right)^{(2)}_{p,q}\left(\frac{e}{2}\right)^2$$

$$+ \frac{1}{3!}\left(\frac{1}{\varepsilon_r}\frac{\partial \varepsilon_r}{\partial x}E_x\right)^{(3)}_{p,q}\left(\frac{e}{2}\right)^3 + \cdots, \quad (4.68)$$

$$\left(\frac{1}{\varepsilon_r}\frac{\partial \varepsilon_r}{\partial x}E_x\right)_{p-1/2,q} = \left(\frac{1}{\varepsilon_r}\frac{\partial \varepsilon_r}{\partial x}E_x\right)_{p,q} - \frac{1}{1!}\left(\frac{1}{\varepsilon_r}\frac{\partial \varepsilon_r}{\partial x}E_x\right)^{(1)}_{p,q}\left(\frac{w}{2}\right)$$

$$+ \frac{1}{2!}\left(\frac{1}{\varepsilon_r}\frac{\partial \varepsilon_r}{\partial x}E_x\right)^{(2)}_{p,q}\left(\frac{w}{2}\right)^2$$

$$- \frac{1}{3!}\left(\frac{1}{\varepsilon_r}\frac{\partial \varepsilon_r}{\partial x}E_x\right)^{(3)}_{p,q}\left(\frac{w}{2}\right)^3 + \cdots. \quad (4.69)$$

Subtracting Eq. (4.69) from Eq. (4.68), we get

$$\left(\frac{1}{\varepsilon_r}\frac{\partial \varepsilon_r}{\partial x}E_x\right)_{p+1/2,q} - \left(\frac{1}{\varepsilon_r}\frac{\partial \varepsilon_r}{\partial x}E_x\right)_{p-1/2,q}$$

$$= \left(\frac{1}{\varepsilon_r}\frac{\partial \varepsilon_r}{\partial x}E_x\right)^{(1)}_{p,q}\left(\frac{e+w}{2}\right) + \left(\frac{1}{\varepsilon_r}\frac{\partial \varepsilon_r}{\partial x}E_x\right)^{(2)}_{p,q}\frac{1}{2!}\frac{(e+w)(e-w)}{4}$$

$$+ \left(\frac{1}{\varepsilon_r}\frac{\partial \varepsilon_r}{\partial x}E_x\right)^{(3)}_{p,q}\frac{1}{3!}\frac{(e+w)(e^2-ew-w^2)}{8} + \cdots.$$

Thus, the first derivative is

$$\frac{\partial}{\partial x}\left(\frac{1}{\varepsilon_r}\frac{\partial \varepsilon_r}{\partial x}E_x\right)_{p,q} = \frac{2}{e+w}\left\{\left(\frac{1}{\varepsilon_r}\frac{\partial \varepsilon_r}{\partial x}E_x\right)_{p+1/2,q} - \left(\frac{1}{\varepsilon_r}\frac{\partial \varepsilon_r}{\partial x}E_x\right)_{p-1/2,q}\right\}$$

$$- \left(\frac{1}{\varepsilon_r}\frac{\partial \varepsilon_r}{\partial x}E_x\right)^{(2)}_{p,q}\frac{e-w}{4}$$

$$- \left(\frac{1}{\varepsilon_r}\frac{\partial \varepsilon_r}{\partial x}E_x\right)^{(3)}_{p,q}\frac{e^2-ew-w^2}{24} + \cdots. \qquad (4.70)$$

Finally, we get

$$\frac{\partial}{\partial x}\left(\frac{1}{\varepsilon_r}\frac{\partial \varepsilon_r}{\partial x}E_x\right)_{p,q} = \frac{2}{e+w}\left\{\left(\frac{1}{\varepsilon_r}\frac{\partial \varepsilon_r}{\partial x}E_x\right)_{p+1/2,q} - \left(\frac{1}{\varepsilon_r}\frac{\partial \varepsilon_r}{\partial x}E_x\right)_{p-1/2,q}\right\}.$$
$$(4.71)$$

According to Eq. (4.70), the error caused by the finite-difference approximation in Eq. (4.71) is $O(e-w)$ when $e \neq w$ (nonequidistant discretization) and is $O((\Delta x)^2)$ when $e = w = \Delta x$ (equidistant discretization).

The next step is to derive the finite-difference expression for Eq. (4.71), and this is done by calculating the two terms within the brackets on the right-hand side.

4.3 FINITE-DIFFERENCE EXPRESSIONS OF WAVE EQUATIONS

Calculating Taylor series expansions of $E_{p+1,q}$ and $E_{p,q}$ around the hypothetical difference center $(p+\frac{1}{2}, q)$, we get

$$E_{p+1,q} = E_{p+1/2,q} + \frac{1}{1!}\frac{\partial E}{\partial x}\bigg|_{p+1/2,q} \cdot \left(\frac{e}{2}\right) + \frac{1}{2!}\frac{\partial^2 E}{\partial x^2}\bigg|_{p+1/2,q} \cdot \left(\frac{e}{2}\right)^2$$
$$+ \frac{1}{3!}\frac{\partial^3 E}{\partial x^3}\bigg|_{p+1/2,q} \cdot \left(\frac{e}{2}\right)^3 + O(e^4), \tag{4.72}$$

$$E_{p,q} = E_{p+1/2,q} - \frac{1}{1!}\frac{\partial E}{\partial x}\bigg|_{p+1/2,q} \cdot \left(\frac{e}{2}\right) + \frac{1}{2!}\frac{\partial^2 E}{\partial x^2}\bigg|_{p+1/2,q} \cdot \left(\frac{e}{2}\right)^2$$
$$- \frac{1}{3!}\frac{\partial^3 E}{\partial x^3}\bigg|_{p+1/2,q} \cdot \left(\frac{e}{2}\right)^3 + O(e^4). \tag{4.73}$$

Adding Eq. (4.72) to Eq. (4.73), we get

$$E_{p+1,q} + E_{p,q} = 2E_{p+1/2,q} + \frac{\partial^2 E}{\partial x^2}\bigg|_{p+1/2,q} \cdot \left(\frac{e}{2}\right)^2 + \frac{\partial^4 E}{\partial x^4}\bigg|_{p+1/2,q} \cdot \frac{1}{12}\left(\frac{e}{2}\right)^4 + \cdots.$$

Therefore

$$E_{p+1/2,q} = \frac{1}{2}(E_{p+1,q} + E_{p,q}) + O(e^2). \tag{4.74}$$

Using the expression $\varepsilon_r(p, q)$ for $\varepsilon_r(x_p, y_q)$ and calculating Taylor series expansions of $\varepsilon_r(p+1, q)$ and $\varepsilon_r(p, q)$ around the hypothetical difference center $(p+\frac{1}{2}, q)$, we get

$$\varepsilon_r(p+1, q) = \varepsilon_r\left(p+\frac{1}{2}, q\right) + \frac{1}{1!}\frac{\partial \varepsilon_r}{\partial x}\bigg|_{p+1/2,q} \cdot \left(\frac{e}{2}\right) + \frac{1}{2!}\frac{\partial^2 \varepsilon_r}{\partial x^2}\bigg|_{p+1/2,q} \cdot \left(\frac{e}{2}\right)^2$$
$$+ \frac{1}{3!}\frac{\partial^3 \varepsilon_r}{\partial x^3}\bigg|_{p+1/2,q} \cdot \left(\frac{e}{2}\right)^3 + O(e^4), \tag{4.75}$$

$$\varepsilon_r(p, q) = \varepsilon_r\left(p+\frac{1}{2}, q\right) - \frac{1}{1!}\frac{\partial \varepsilon_r}{\partial x}\bigg|_{p+1/2,q} \cdot \left(\frac{e}{2}\right) + \frac{1}{2!}\frac{\partial^2 \varepsilon_r}{\partial x^2}\bigg|_{p+1/2,q} \cdot \left(\frac{e}{2}\right)^2$$
$$- \frac{1}{3!}\frac{\partial^3 \varepsilon_r}{\partial x^3}\bigg|_{p+1/2,q} \cdot \left(\frac{e}{2}\right)^3 + O(e^4). \tag{4.76}$$

And adding Eq. (4.75) to Eq. (4.76), we get

$$\varepsilon_r(p+1,q) + \varepsilon_r(p,q) = 2\varepsilon_r\left(p+\frac{1}{2},q\right) + \left.\frac{\partial^2 \varepsilon_r}{\partial x^2}\right|_{p+1/2,q} \cdot \left(\frac{e}{2}\right)^2$$
$$+ \left.\frac{\partial^4 \varepsilon_r}{\partial x^4}\right|_{p+1/2,q} \cdot \frac{1}{12}\left(\frac{e}{2}\right)^4 + \cdots.$$

Therefore

$$\varepsilon_r(p+\tfrac{1}{2},q) = \tfrac{1}{2}\{\varepsilon_r(p+1,q) + \varepsilon_r(p,q)\} + O(e^2). \qquad (4.77)$$

On the other hand, subtracting Eq. (4.76) from Eq. (4.75), we get

$$\varepsilon_r(p+1,q) - \varepsilon_r(p,q) = \left.\frac{\partial \varepsilon_r}{\partial x}\right|_{p+1/2,q} \cdot e + \left.\frac{\partial^3 \varepsilon_r}{\partial x^3}\right|_{p+1/2,q} \cdot \frac{2}{3!}\left(\frac{e}{2}\right)^3 + \cdots.$$

Thus

$$\left.\frac{\partial \varepsilon_r}{\partial x}\right|_{p+1/2,q} = \frac{1}{e}\{\varepsilon_r(p+1,q) - \varepsilon_r(p,q)\} + O(e^2). \qquad (4.78)$$

Combining Eqs. (4.77) and (4.78), we get

$$\frac{1}{\varepsilon_r}\left.\frac{\partial \varepsilon_r}{\partial x}\right|_{p+1/2,q} = \left[\frac{1}{2}\{\varepsilon_r(p+1,q) + \varepsilon_r(p,q)\} + O(e^2)\right]^{-1}$$
$$\times \left[\frac{1}{e}\{\varepsilon_r(p+1,q) - \varepsilon_r(p,q)\} + O(e^2)\right]$$
$$= \left\{\frac{2}{\varepsilon_r(p+1,q) + \varepsilon_r(p,q)} + O(e^2)\right\}$$
$$\times \left[\frac{1}{e}\{\varepsilon_r(p+1,q) - \varepsilon_r(p,q)\} + O(e^2)\right]$$
$$= \frac{2}{e}\frac{\varepsilon_r(p+1,q) - \varepsilon_r(p,q)}{\varepsilon_r(p+1,q) + \varepsilon_r(p,q)} + O(e). \qquad (4.79)$$

4.3 FINITE-DIFFERENCE EXPRESSIONS OF WAVE EQUATIONS

Finally, we can derive the following equation by using Eqs. (4.74) and (4.79):

$$\frac{1}{\varepsilon_r}\frac{\partial \varepsilon_r}{\partial x}E\bigg|_{p+1/2,q} = \left\{\frac{2}{e}\frac{\varepsilon_r(p+1,q)-\varepsilon_r(p,q)}{\varepsilon_r(p+1,q)+\varepsilon_r(p,q)} + O(e)\right\}$$
$$\times \left(\frac{E_{p+1,q}+E_{p,q}}{2} + O(e^2)\right)$$
$$= \frac{1}{e}\frac{\varepsilon_r(p+1,q)-\varepsilon_r(p,q)}{\varepsilon_r(p+1,q)+\varepsilon_r(p,q)}(E_{p+1,q}+E_{p,q}) + O(e). \tag{4.80}$$

Now that we have the first term within the brackets on the right-hand side of Eq. (4.71), we can derive the second term by using the same procedure. Calculating Taylor series expansions of $E_{p,q}$ and $E_{p-1,q}$ around the hypothetical difference center ($p - \frac{1}{2}, q$), we get

$$E_{p,q} = E_{p-1/2,q} + \frac{1}{1!}\frac{\partial E}{\partial x}\bigg|_{p-1/2,q}\cdot\left(\frac{w}{2}\right) + \frac{1}{2!}\frac{\partial^2 E}{\partial x^2}\bigg|_{p-1/2,q}\cdot\left(\frac{w}{2}\right)^2$$
$$+ \frac{1}{3!}\frac{\partial^3 E}{\partial x^3}\bigg|_{p-1/2,q}\cdot\left(\frac{w}{2}\right)^3 + O(w^4), \tag{4.81}$$

$$E_{p-1,q} = E_{p-1/2,q} - \frac{1}{1!}\frac{\partial E}{\partial x}\bigg|_{p-1/2,q}\cdot\left(\frac{w}{2}\right) + \frac{1}{2!}\frac{\partial^2 E}{\partial x^2}\bigg|_{p-1/2,q}\cdot\left(\frac{w}{2}\right)^2$$
$$- \frac{1}{3!}\frac{\partial^3 E}{\partial x^3}\bigg|_{p-1/2,q}\cdot\left(\frac{w}{2}\right)^3 + O(w^4). \tag{4.82}$$

Adding Eq. (4.81) to Eq. (4.82), we get

$$E_{p-1,q} + E_{p,q} = 2E_{p-1/2,q} + \frac{\partial^2 E}{\partial x^2}\bigg|_{p-1/2,q}\cdot\left(\frac{w}{2}\right)^2$$
$$+ \frac{\partial^4 E}{\partial x^4}\bigg|_{p-1/2,q}\cdot\frac{1}{12}\left(\frac{w}{2}\right)^4 + \cdots.$$

Therefore

$$E_{p-1/2,q} = \tfrac{1}{2}(E_{p,q} + E_{p-1,q}) + O(w^2). \tag{4.83}$$

Similarly, calculating Taylor series expansions of $\varepsilon_r(p-1, q)$ and $\varepsilon_r(p, q)$ around the hypothetical difference center $(p - \frac{1}{2}, q)$, we get

$$\varepsilon_r(p-1, q) = \varepsilon_r\left(p - \frac{1}{2}, q\right) - \frac{1}{1!} \left.\frac{\partial \varepsilon_r}{\partial x}\right|_{p-1/2, q} \cdot \left(\frac{w}{2}\right) + \frac{1}{2!} \left.\frac{\partial^2 \varepsilon_r}{\partial x^2}\right|_{p-1/2, q}$$
$$\times \left(\frac{w}{2}\right)^2 - \frac{1}{3!} \left.\frac{\partial^3 \varepsilon_r}{\partial x^3}\right|_{p-1/2, q} \cdot \left(\frac{w}{2}\right)^3 + O(w^4), \quad (4.84)$$

$$\varepsilon_r(p, q) = \varepsilon_r\left(p - \frac{1}{2}, q\right) + \frac{1}{1!} \left.\frac{\partial \varepsilon_r}{\partial x}\right|_{p-1/2, q} \cdot \left(\frac{w}{2}\right) + \frac{1}{2!} \left.\frac{\partial^2 \varepsilon_r}{\partial x^2}\right|_{p-1/2, q}$$
$$\times \left(\frac{w}{2}\right)^2 + \frac{1}{3!} \left.\frac{\partial^3 \varepsilon_r}{\partial x^3}\right|_{p-1/2, q} \cdot \left(\frac{w}{2}\right)^3 + O(w^4). \quad (4.85)$$

Adding Eq. (4.84) to Eq. (4.85) gives

$$\varepsilon_r(p-1, q) + \varepsilon_r(p, q) = 2\varepsilon_r(p - \tfrac{1}{2}, q) + \left.\frac{\partial^2 \varepsilon_r}{\partial x^2}\right|_{p-1/2, q} \cdot \left(\frac{w}{2}\right)^2$$
$$+ \left.\frac{\partial^4 \varepsilon_r}{\partial x^4}\right|_{p-1/2, q} \cdot \frac{1}{12}\left(\frac{w}{2}\right)^4 + \cdots,$$
$$\therefore \quad \varepsilon_r(p - \tfrac{1}{2}, q) = \tfrac{1}{2}\{\varepsilon_r(p-1, q) + \varepsilon_r(p, q)\} + O(w^2), \quad (4.86)$$

whereas subtracting Eq. (4.85) from Eq. (4.86) gives

$$\varepsilon_r(p, q) - \varepsilon_r(p-1, q) = \left.\frac{\partial \varepsilon_r}{\partial x}\right|_{p-1/2, q} \cdot w + \left.\frac{\partial^3 \varepsilon_r}{\partial x^3}\right|_{p-1/2, q} \cdot \frac{2}{3!}\left(\frac{w}{2}\right)^3 + \cdots,$$
$$\therefore \quad \left.\frac{\partial \varepsilon_r}{\partial x}\right|_{p-1/2, q} = \frac{1}{w}\{\varepsilon_r(p, q) - \varepsilon_r(p-1, q)\} + O(w^2).$$
$$(4.87)$$

4.3 FINITE-DIFFERENCE EXPRESSIONS OF WAVE EQUATIONS

Combining Eqs. (4.86) and (4.87), we get

$$\left.\frac{1}{\varepsilon_r}\frac{\partial \varepsilon_r}{\partial x}\right|_{p-1/2,q} = \left[\frac{1}{2}\{\varepsilon_r(p-1,q)+\varepsilon_r(p,q)\}+O(w^2)\right]^{-1}$$
$$\times \left[\frac{1}{w}\{\varepsilon_r(p,q)-\varepsilon_r(p-1,q)\}+O(w^2)\right]$$
$$= \left\{\frac{2}{\varepsilon_r(p,q)+\varepsilon_r(p-1,q)}+O(w^2)\right\}$$
$$\times \left[\frac{1}{e}\{\varepsilon_r(p,q)-\varepsilon_r(p-1,q)\}+O(w^2)\right]$$
$$= \frac{2}{e}\frac{\varepsilon_r(p,q)-\varepsilon_r(p-1,q)}{\varepsilon_r(p,q)+\varepsilon_r(p-1,q)}+O(w). \qquad (4.88)$$

Finally, we can derive the following equation by using Eqs. (4.83) and (4.88):

$$\left.\frac{1}{\varepsilon_r}\frac{\partial \varepsilon_r}{\partial x}E\right|_{p-1/2,q} = \left\{\frac{2}{w}\frac{\varepsilon_r(p,q)-\varepsilon_r(p-1,q)}{\varepsilon_r(p,q)+\varepsilon_r(p-1,q)}+O(w)\right\}$$
$$\times \left(\frac{E_{p,q}+E_{p-1,q}}{2}+O(w^2)\right)$$
$$= \frac{1}{w}\frac{\varepsilon_r(p,q)-\varepsilon_r(p-1,q)}{\varepsilon_r(p,q)+\varepsilon_r(p-1,q)}(E_{p,q}+E_{p-1,q})+O(w). \qquad (4.89)$$

Now that we have obtained the two terms within the brackets on the right-hand side of Eq. (4.71) by using Taylor series expansions, let us further our understanding by deriving equations for these terms without using Taylor series expansions. With respect to the first term, the difference center is $(p+\frac{1}{2},q)$. To set the difference center at $(p+\frac{1}{2},q)$, we need to take the average of $(p+1,q)$ and (p,q). Thus, we get the relations

$$E_{p+1/2,q} = \tfrac{1}{2}(E_{p+1,q}+E_{p,q}), \qquad (4.90)$$

$$\varepsilon_r(p+\tfrac{1}{2},q) = \tfrac{1}{2}\{\varepsilon_r(p+1,q)+\varepsilon_r(p,q)\}, \qquad (4.91)$$

$$\left.\frac{\partial \varepsilon_r}{\partial x}\right|_{p+1/2,q} = \frac{1}{e}\{\varepsilon_r(p+1,q)-\varepsilon_r(p,q)\}. \qquad (4.92)$$

We can immediately derive the equation

$$\left.\frac{1}{\varepsilon_r}\frac{\partial \varepsilon_r}{\partial x}E\right|_{p+1/2,q} = -\frac{1}{e}\frac{\varepsilon_r(p+1,q) - \varepsilon_r(p,q)}{\varepsilon_r(p+1,q) + \varepsilon_r(p,q)}(E_{p+1,q} + E_{p,q}). \quad (4.93)$$

With respect to the second term, on the other hand, the hypothetical difference center is $(p-\frac{1}{2}, q)$. To set the difference center at $(p-\frac{1}{2}, q)$, we need to take the average of (p, q) and $(p-1, q)$. Thus, we get the relations

$$E_{p-1/2,q} = \tfrac{1}{2}(E_{p,q} + E_{p-1,q}), \quad (4.94)$$

$$\varepsilon_r(p-\tfrac{1}{2}, q) = \tfrac{1}{2}\{\varepsilon_r(p, q) + \varepsilon_r(p-1, q)\}, \quad (4.95)$$

$$\left.\frac{\partial \varepsilon_r}{\partial x}\right|_{p-1/2,q} = \frac{1}{w}\{\varepsilon_r(p, q) - \varepsilon_r(p-1, q)\}. \quad (4.96)$$

Again we can immediately derive the equation

$$\left.\frac{1}{\varepsilon_r}\frac{\partial \varepsilon_r}{\partial x}E\right|_{p-1/2,q} = \frac{1}{w}\frac{\varepsilon_r(p,q) - \varepsilon_r(p-1,q)}{\varepsilon_r(p,q) + \varepsilon_r(p-1,q)}(E_{p,q} + E_{p-1,q}). \quad (4.97)$$

We can see that Eqs. (4.93) and (4.97) are equivalent to Eqs. (4.80) and (4.89).

Let us return here to the main topic. Substituting Eqs. (4.80) and (4.89) into Eq. (4.71), we get

$$\frac{\partial}{\partial x}\left(\frac{1}{\varepsilon_r}\frac{\partial \varepsilon_r}{\partial x}E\right) = \frac{2}{e+w}\left\{\frac{1}{e}\frac{\varepsilon_r(p+1,q) - \varepsilon_r(p,q)}{\varepsilon_r(p+1,q) + \varepsilon_r(p,q)}(E_{p+1,q} + E_{p,q})\right.$$

$$\left. - \frac{1}{w}\frac{\varepsilon_r(p,q) - \varepsilon_r(p-1,q)}{\varepsilon_r(p,q) + \varepsilon_r(p-1,q)}(E_{p,q} + E_{p-1,q})\right\}. \quad (4.98)$$

4.3 FINITE-DIFFERENCE EXPRESSIONS OF WAVE EQUATIONS

Substituting Eqs. (4.65), (4.67), and (4.98) into Eq. (4.50), we get

$$\frac{2}{e(e+w)}E_{p+1,q} + \frac{2}{w(e+w)}E_{p-1,q} - \frac{2}{ew}E_{p,q} + \frac{2}{e+w}$$

$$\times \left\{ \frac{1}{e} \frac{\varepsilon_r(p+1,q) - \varepsilon_r(p,q)}{\varepsilon_r(p+1,q) + \varepsilon_r(p,q)}(E_{p+1,q} + E_{p,q}) \right.$$

$$\left. - \frac{1}{w} \frac{\varepsilon_r(p,q) - \varepsilon_r(p-1,q)}{\varepsilon_r(p,q) + \varepsilon_r(p-1,q)}(E_{p,q} + E_{p-1,q}) \right\}$$

$$+ \frac{2}{s(s+n)}E_{p,q+1} + \frac{2}{n(s+n)}E_{p,q-1} - \frac{2}{ns}E_{p,q} + \{k_0^2 \varepsilon_r(p,q) - \beta^2\}E_{p,q}$$

$$= \frac{2}{w(e+w)} \left\{ 1 - \frac{\varepsilon_r(p,q) - \varepsilon_r(p-1,q)}{\varepsilon_r(p,q) + \varepsilon_r(p-1,q)} \right\} E_{p-1,q}$$

$$+ \frac{2}{e(e+w)} \left\{ 1 - \frac{\varepsilon_r(p,q) - \varepsilon_r(p+1,q)}{\varepsilon_r(p,q) + \varepsilon_r(p+1,q)} \right\} E_{p+1,q}$$

$$+ \left\{ -\frac{2}{ew} - \frac{2}{w(e+w)} \frac{\varepsilon_r(p,q) - \varepsilon_r(p-1,q)}{\varepsilon_r(p,q) + \varepsilon_r(p-1,q)} \right.$$

$$\left. - \frac{2}{e(e+w)} \frac{\varepsilon_r(p,q) - \varepsilon_r(p+1,q)}{\varepsilon_r(p,q) + \varepsilon_r(p+1,q)} \right\} E_{p,q}$$

$$+ \frac{2}{n(s+n)}E_{p,q-1} + \frac{2}{s(s+n)}E_{p,q+1} - \frac{1}{ns}E_{p,q} + \{k_0^2\varepsilon_r(p,q) - \beta^2\}E_{p,q}$$

$$= \frac{2}{w(e+w)} \frac{2\varepsilon_r(p-1,q)}{\varepsilon_r(p,q) + \varepsilon_r(p-1,q)} E_{p-1,q}$$

$$+ \frac{2}{e(e+w)} \frac{2\varepsilon_r(p+1,q)}{\varepsilon_r(p,q) + \varepsilon_r(p+1,q)} E_{p+1,q}$$

$$+ \left\{ -\frac{2}{ew} - \frac{2}{w(e+w)} \frac{\varepsilon_r(p,q) - \varepsilon_r(p-1,q)}{\varepsilon_r(p,q) + \varepsilon_r(p-1,q)} \right.$$

$$\left. - \frac{2}{e(e+w)} \frac{\varepsilon_r(p,q) - \varepsilon_r(p+1,q)}{\varepsilon_r(p,q) + \varepsilon_r(p+1,q)} \right\} E_{p,q}$$

$$+ \frac{2}{n(s+n)}E_{p,q-1} + \frac{2}{s(s+n)}E_{p,q+1} - \frac{1}{ns}E_{p,q}$$

$$+ \{k_0^2 \varepsilon_r(p,q) - \beta^2\}E_{p,q} = 0.$$

Thus, we get the following finite difference expression for Eq. (4.50):

$$\alpha_w E_{p-1,q} + \alpha_e E_{p+1,q} + \alpha_n E_{p,q-1} + \alpha_s E_{p,q+1}$$
$$+ (\alpha_x + \alpha_y)E_{p,q} + \{k_0^2 \varepsilon_r(p,q) - \beta^2\}E_{p,q} = 0, \quad (4.99)$$

where

$$\alpha_w = \frac{2}{w(e+w)} \frac{2\varepsilon_r(p-1,q)}{\varepsilon_r(p,q) + \varepsilon_r(p-1,q)}, \quad (4.100)$$

$$\alpha_e = \frac{2}{e(e+w)} \frac{2\varepsilon_r(p+1,q)}{\varepsilon_r(p,q) + \varepsilon_r(p+1,q)}, \quad (4.101)$$

$$\alpha_n = \frac{2}{n(n+s)}, \quad (4.102)$$

$$\alpha_s = \frac{2}{s(n+s)}, \quad (4.103)$$

$$\alpha_x = -\frac{2}{ew} - \frac{2}{w(e+w)} \frac{\varepsilon_r(p,q) - \varepsilon_r(p-1,q)}{\varepsilon_r(p,q) + \varepsilon_r(p-1,q)}$$
$$- \frac{2}{e(e+w)} \frac{\varepsilon_r(p,q) - \varepsilon_r(p+1,q)}{\varepsilon_r(p,q) + \varepsilon_r(p+1,q)}$$
$$= -\frac{4}{ew} + \alpha_e + \alpha_w, \quad (4.104)$$

$$\alpha_y = -\frac{2}{ns} = -\alpha_n - \alpha_s. \quad (4.105)$$

B. H_y Representation The H_y representation wave equation derived for the quasi-TE mode [Eq. (4.43)] is

$$\varepsilon_r \frac{\partial}{\partial x}\left(\frac{1}{\varepsilon_r}\frac{\partial H_y}{\partial x}\right) + \frac{\partial^2 H_y}{\partial y^2} + (k_0^2 \varepsilon_r - \beta^2)H_y = 0. \quad (4.106)$$

Now, let us derive the finite-difference expression for this equation. To do this, we start with the first term of Eq. (4.106). Again using $(p+\frac{1}{2}, q)$ as the hypothetical finite-difference center between nodes (p,q) and

4.3 FINITE-DIFFERENCE EXPRESSIONS OF WAVE EQUATIONS

$(p+1, q)$ and using $(p - \frac{1}{2}, q)$ as the hypothetical finite-difference center between nodes $(p-1, q)$ and (p, q), we get

$$\left(\frac{1}{\varepsilon_r}\frac{\partial H_y}{\partial x}\right)_{p+1/2,q} = \left(\frac{1}{\varepsilon_r}\frac{\partial H_y}{\partial x}\right)_{p,q} + \frac{1}{1!}\left(\frac{1}{\varepsilon_r}\frac{\partial H_y}{\partial x}\right)^{(1)}_{p,q}\left(\frac{e}{2}\right)$$
$$+ \frac{1}{2!}\left(\frac{1}{\varepsilon_r}\frac{\partial H_y}{\partial x}\right)^{(2)}_{p,q}\left(\frac{e}{2}\right)^2$$
$$+ \frac{1}{3!}\left(\frac{1}{\varepsilon_r}\frac{\partial H_y}{\partial x}\right)^{(3)}_{p,q}\left(\frac{e}{2}\right)^3 + \cdots, \quad (4.107)$$

$$\left(\frac{1}{\varepsilon_r}\frac{\partial H_y}{\partial x}\right)_{p-1/2,q} = \left(\frac{1}{\varepsilon_r}\frac{\partial H_y}{\partial x}\right)_{p,q} - \frac{1}{1!}\left(\frac{1}{\varepsilon_r}\frac{\partial H_y}{\partial x}\right)^{(1)}_{p,q}\left(\frac{w}{2}\right)$$
$$+ \frac{1}{2!}\left(\frac{1}{\varepsilon_r}\frac{\partial H_y}{\partial x}\right)^{(2)}_{p,q}\left(\frac{w}{2}\right)^2$$
$$- \frac{1}{3!}\left(\frac{1}{\varepsilon_r}\frac{\partial H_y}{\partial x}\right)^{(3)}_{p,q}\left(\frac{w}{2}\right)^3 + \cdots. \quad (4.108)$$

Subtracting Eq. (4.107) from Eq. (4.106), we get

$$\left(\frac{1}{\varepsilon_r}\frac{\partial H_y}{\partial x}\right)_{p+1/2,q} - \left(\frac{1}{\varepsilon_r}\frac{\partial H_y}{\partial x}\right)_{p-1/2,q}$$
$$= \left(\frac{1}{\varepsilon_r}\frac{\partial H_y}{\partial x}\right)^{(1)}_{p,q}\left(\frac{e+w}{2}\right)$$
$$+ \left(\frac{1}{\varepsilon_r}\frac{\partial H_y}{\partial x}\right)^{(2)}_{p,q}\frac{1}{2!}\frac{(e+w)(e-w)}{4}$$
$$+ \left(\frac{1}{\varepsilon_r}\frac{\partial H_y}{\partial x}\right)^{(3)}_{p,q}\frac{1}{3!}\frac{(e+w)(e^2 - ew - w^2)}{8} + \cdots.$$

Thus,

$$\frac{\partial}{\partial x}\left(\frac{1}{\varepsilon_r}\frac{\partial H_y}{\partial x}\right)_{p,q} = \frac{2}{e+w}\left\{\left(\frac{1}{\varepsilon_r}\frac{\partial H_y}{\partial x}\right)_{p+1/2,q} - \left(\frac{1}{\varepsilon_r}\frac{\partial H_y}{\partial x}\right)_{p-1/2,q}\right\}$$
$$- \left(\frac{1}{\varepsilon_r}\frac{\partial H_y}{\partial x}\right)^{(2)}_{p,q}\frac{e-w}{4} - \left(\frac{1}{\varepsilon_r}\frac{\partial H_y}{\partial x}\right)^{(3)}_{p,q}\frac{e^2 - ew - w^2}{24} + \cdots.$$
$$(4.109)$$

Finally, we obtain

$$\frac{\partial}{\partial x}\left(\frac{1}{\varepsilon_r}\frac{\partial H_y}{\partial x}\right)_{p,q} = \frac{2}{e+w}\left\{\left(\frac{1}{\varepsilon_r}\frac{\partial H_y}{\partial x}\right)_{p+1/2,q} - \left(\frac{1}{\varepsilon_r}\frac{\partial H_y}{\partial x}\right)_{p-1/2,q}\right\}. \quad (4.110)$$

According to Eq. (4.109), the error caused by the finite-difference approximation in Eq. (4.110) is $O(e - w)$ when $e \neq w$ (nonequidistant discretization) and is $O((\Delta x)^2)$ when $e = w = \Delta x$ (equidistant discretization).

The next step is to derive the finite-difference expression for Eq. (4.110) by calculating the two terms within the brackets on the right-hand side.

Now, we calculate the first term. Calculating Taylor series expansions of $H_{p+1,q}$ and $H_{p,q}$ around the hypothetical difference center $(p+\frac{1}{2}, q)$, we get

$$H_{p+1,q} = H_{p+1/2,q} + \frac{1}{1!}\left.\frac{\partial H}{\partial x}\right|_{p+1/2,q} \cdot \left(\frac{e}{2}\right) + \frac{1}{2!}\left.\frac{\partial^2 H}{\partial x^2}\right|_{p+1/2,q} \cdot \left(\frac{e}{2}\right)^2$$

$$+ \frac{1}{3!}\left.\frac{\partial^3 H}{\partial x^3}\right|_{p+1/2,q} \cdot \left(\frac{e}{2}\right)^3 + O(e^4), \quad (4.111)$$

$$H_{p,q} = H_{p+1/2,q} - \frac{1}{1!}\left.\frac{\partial H}{\partial x}\right|_{p+1/2,q} \cdot \left(\frac{e}{2}\right) + \frac{1}{2!}\left.\frac{\partial^2 H}{\partial x^2}\right|_{p+1/2,q} \cdot \left(\frac{e}{2}\right)^2$$

$$- \frac{1}{3!}\left.\frac{\partial^3 H}{\partial x^3}\right|_{p+1/2,q} \cdot \left(\frac{e}{2}\right)^3 + O(e^4). \quad (4.112)$$

Subtracting Eq. (4.112) from Eq. (4.111), we get

$$H_{p+1/2,q} - H_{p,q} = \left.\frac{\partial H}{\partial x}\right|_{p+1/2,q} \cdot e + \left.\frac{\partial^3 H}{\partial x^3}\right|_{p+1/2,q} \cdot \frac{e^3}{24} + \cdots.$$

Therefore

$$\left.\frac{\partial H}{\partial x}\right|_{p+1/2,q} = \frac{1}{e}(H_{p+1,q} - H_{p,q}) + O(e^2). \quad (4.113)$$

4.3 FINITE-DIFFERENCE EXPRESSIONS OF WAVE EQUATIONS

Recall that the relative permittivity at $(p+\frac{1}{2}, q)$ [Eq. (4.77)] is

$$\varepsilon_r(p+\tfrac{1}{2}, q) = \tfrac{1}{2}\{\varepsilon_r(p+1, q) + \varepsilon_r(p, q)\} + O(e^2). \tag{4.114}$$

Thus, according to Eqs. (4.113) and (4.114),

$$\begin{aligned}
\frac{1}{\varepsilon_r}\frac{\partial H_y}{\partial x}\bigg|_{p+1/2,q} &= [\tfrac{1}{2}\{\varepsilon_r(p+1,q)+\varepsilon_r(p,q)\}+O(e^2)]^{-1} \\
&\quad \times \left[\frac{1}{e}(H_{p+1,q}-H_{p,q})+O(e^2)\right] \\
&= \left\{\frac{2}{\varepsilon_r(p+1,q)+\varepsilon_r(p,q)}+O(e^2)\right\} \\
&\quad \times \left[\frac{1}{e}(H_{p+1,q}-H_{p,q})+O(e^2)\right] \\
&= \frac{1}{e}\frac{2}{\varepsilon_r(p+1,q)+\varepsilon_r(p,q)}(H_{p+1,q}-H_{p,q})+O(e).
\end{aligned}$$
(4.115)

Next, we calculate the second term within the brackets in Eq. (4.110). Calculating Taylor series expansions of $H_{p,q}$ and $H_{p-1,q}$ around the hypothetical difference center $(p-\frac{1}{2}, q)$, we get

$$H_{p,q} = H_{p-1/2,q} + \frac{1}{1!}\frac{\partial H}{\partial x}\bigg|_{p-1/2,q}\cdot\left(\frac{w}{2}\right) + \frac{1}{2!}\frac{\partial^2 H}{\partial x^2}\bigg|_{p-1/2,q}\cdot\left(\frac{w}{2}\right)^2$$

$$+ \frac{1}{3!}\frac{\partial^3 H}{\partial x^3}\bigg|_{p-1/2,q}\cdot\left(\frac{w}{2}\right)^3 + O(w^4), \tag{4.116}$$

$$H_{p-1,q} = H_{p-1/2,q} - \frac{1}{1!}\frac{\partial H}{\partial x}\bigg|_{p-1/2,q}\cdot\left(\frac{w}{2}\right) + \frac{1}{2!}\frac{\partial^2 H}{\partial x^2}\bigg|_{p-1/2,q}\cdot\left(\frac{w}{2}\right)^2$$

$$- \frac{1}{3!}\frac{\partial^3 H}{\partial x^3}\bigg|_{p-1/2,q}\cdot\left(\frac{w}{2}\right)^3 + O(w^4). \tag{4.117}$$

Subtracting Eq. (4.117) from Eq. (4.116), we get

$$H_{p,q} - H_{p-1,q} = \left.\frac{\partial H}{\partial x}\right|_{p-1/2,q} \cdot w + \left.\frac{\partial^3 H}{\partial x^3}\right|_{p-1/2,q} \cdot \frac{w^3}{24} + \cdots.$$

Therefore

$$\left.\frac{\partial H}{\partial x}\right|_{p-1/2,q} = \frac{1}{w}(H_{p,q} - H_{p-1,q}) + O(w^2). \tag{4.118}$$

Recalling that the relative permittivity at $(p - \frac{1}{2}, q)$ [Eq. (4.86)] is

$$\varepsilon_r(p - \tfrac{1}{2}, q) = \tfrac{1}{2}\{\varepsilon_r(p, q) + \varepsilon_r(p - 1, q)\} + O(w^2) \tag{4.119}$$

and using Eqs. (4.118) and (4.119), we get

$$\frac{1}{\varepsilon_r}\left.\frac{\partial H_y}{\partial x}\right|_{p-1/2,q} = [\tfrac{1}{2}\{\varepsilon_r(p, q) + \varepsilon_r(p - 1, q)\} + O(w^2)]^{-1}$$

$$\times \left[\frac{1}{w}(H_{p,q} - H_{p-1,q}) + O(w^2)\right]$$

$$= \left\{\frac{2}{\varepsilon_r(p, q) + \varepsilon_r(p - 1, q)} + O(w^2)\right\}$$

$$\times \left[\frac{1}{w}(H_{p,q} - H_{p-1,q}) + O(w^2)\right]$$

$$= \frac{1}{w}\frac{2}{\varepsilon_r(p, q) + \varepsilon_r(p - 1, q)}(H_{p,q} - H_{p-1,q}) + O(w).$$

$$\tag{4.120}$$

We now have the finite-difference expressions for the first and second terms of Eq. (4.110). Although it is not shown here, these finite-difference expressions can be derived more easily with the procedures used in deriving Eqs. (4.93) and (4.97) (see Problem 3).

4.3 FINITE-DIFFERENCE EXPRESSIONS OF WAVE EQUATIONS

Substitution of Eqs. (4.115) and (4.120) into Eq. (4.110) results in

$$\frac{\partial}{\partial x}\left(\frac{1}{\varepsilon_r}\frac{\partial H_y}{\partial x}\right) = \frac{2}{e+w}\left\{\frac{1}{e}\frac{2}{\varepsilon_r(p+1,q)+\varepsilon_r(p,q)}(H_{p+1,q}-H_{p,q})\right.$$

$$\left.-\frac{1}{w}\frac{2}{\varepsilon_r(p,q)+\varepsilon_r(p-1,q)}(H_{p,q}-H_{p-1,q})\right\}.$$

Thus, we get the finite-difference expression

$$\varepsilon_r\frac{\partial}{\partial x}\left(\frac{1}{\varepsilon_r}\frac{\partial H_y}{\partial x}\right) = \frac{2}{e+w}\left\{\frac{1}{e}\frac{2\varepsilon_r(p,q)}{\varepsilon_r(p+1,q)+\varepsilon_r(p,q)}(H_{p+1,q}-H_{p,q})\right.$$

$$\left.-\frac{1}{w}\frac{2\varepsilon_r(p,q)}{\varepsilon_r(p,q)+\varepsilon_r(p-1,q)}(H_{p,q}-H_{p-1,q})\right\}$$

$$=\frac{2}{e(e+w)}\frac{2\varepsilon_r(p,q)}{\varepsilon_r(p,q)+\varepsilon_r(p+1,q)}H_{p+1,q}$$

$$-\left\{\frac{2}{e(e+w)}\frac{2\varepsilon_r(p,q)}{\varepsilon_r(p,q)+\varepsilon_r(p+1,q)}\right.$$

$$\left.+\frac{2}{w(e+w)}\frac{2\varepsilon_r(p,q)}{\varepsilon_r(p,q)+\varepsilon_r(p-1,q)}\right\}H_{p,q}$$

$$+\frac{2}{w(e+w)}\frac{2\varepsilon_r(p,q)}{\varepsilon_r(p,q)+\varepsilon_r(p-1,q)}H_{p-1,q}. \quad (4.121)$$

On the other hand, the finite-difference expression for the derivative with respect to y [Eq. (4.67)] is

$$\left.\frac{\partial^2 H}{\partial y^2}\right|_{p,q} = \frac{2}{s(n+s)}H_{p,q+1} + \frac{2}{n(n+s)}H_{p,q-1} - \frac{2}{sn}H_{p,q}. \quad (4.122)$$

Here, the error caused by the finite-difference approximation in Eq. (4.122) is $O(s-n)$ when $n \neq s$ (nonequidistant discretization) and is $O((\Delta y)^2)$ when $s = n = \Delta y$ (equidistant discretization).

Substituting Eqs. (4.121) and (4.122) into Eq. (4.106), we get the following finite-difference expressions for Eq. (4.106):

$$\frac{2}{e+w}\left\{\frac{1}{e}\frac{2\varepsilon_r(p,q)}{\varepsilon_r(p+1,q)+\varepsilon_r(p,q)}(H_{p+1,q}-H_{p,q})\right.$$

$$\left.-\frac{1}{w}\frac{2\varepsilon_r(p,q)}{\varepsilon_r(p,q)+\varepsilon_r(p-1,q)}(H_{p,q}-H_{p-1,q})\right\}$$

$$+\frac{2}{s(n+s)}H_{p,q+1}+\frac{2}{n(n+s)}H_{p,q-1}-\frac{2}{ns}H_{p,q}+\{k_0^2\varepsilon_r(p,q)-\beta^2\}H_{p,q}$$

$$=\frac{2}{e(e+w)}\frac{2\varepsilon_r(p,q)}{\varepsilon_r(p,q)+\varepsilon_r(p+1,q)}H_{p+1,q}$$

$$+\left\{-\frac{2}{ns}-\frac{2}{e(e+w)}\frac{2\varepsilon_r(p,q)}{\varepsilon_r(p,q)+\varepsilon_r(p+1,q)}\right.$$

$$\left.-\frac{2}{w(e+w)}\frac{2\varepsilon_r(p,q)}{\varepsilon_r(p,q)+\varepsilon_r(p-1,q)}\right\}H_{p,q}$$

$$+\frac{2}{w(e+w)}\frac{2\varepsilon_r(p,q)}{\varepsilon_r(p,q)+\varepsilon_r(p-1,q)}H_{p-1,q}$$

$$+\frac{2}{s(n+s)}H_{p,q+1}+\frac{2}{n(n+s)}H_{p,q-1}+\{k_0^2\varepsilon_r(p,q)-\beta^2\}H_{p,q}$$

$$=0$$

Thus, we get the following final finite-difference expression for Eq. (4.106):

$$\alpha_w H_{p-1,q}+\alpha_e H_{p+1,q}+\alpha_n H_{p,q-1}+\alpha_s H_{p,q+1}$$

$$+(\alpha_x+\alpha_y)H_{p,q}+\{k_0^2\varepsilon_r(p,q)-\beta^2\}H_{p,q}=0, \quad (4.123)$$

where

$$\alpha_w = \frac{2}{w(e+w)} \frac{2\varepsilon_r(p,q)}{\varepsilon_r(p,q) + \varepsilon_r(p-1,q)}, \quad (4.124)$$

$$\alpha_e = \frac{2}{e(e+w)} \frac{2\varepsilon_r(p,q)}{\varepsilon_r(p,q) + \varepsilon_r(p+1,q)}, \quad (4.125)$$

$$\alpha_n = \frac{2}{n(n+s)}, \quad (4.126)$$

$$\alpha_s = \frac{2}{s(n+s)}, \quad (4.127)$$

$$\alpha_x = -\frac{2}{w(e+w)} \frac{2\varepsilon_r(p,q)}{\varepsilon_r(p,q) + \varepsilon_r(p-1,q)}$$
$$\quad - \frac{2}{e(e+w)} \frac{2\varepsilon_r(p,q)}{\varepsilon_r(p,q) + \varepsilon_r(p+1,q)}$$
$$= -\alpha_e - \alpha_w, \quad (4.128)$$

$$\alpha_y = -\frac{2}{ns} = -\alpha_n - \alpha_s. \quad (4.129)$$

It should be noted that since the forms of the derivatives with respect to x differ slightly between the wave equations of the electric field and magnetic field representations, the resultant finite-difference expressions for α_w, α_e, and α_x also differ between the two representations.

4.3.2 Quasi-TM Mode

A. E_y Representation The E_y representation wave equation for the quasi-TM mode [Eq. (4.44)] is

$$\frac{\partial^2 E_y}{\partial x^2} + \frac{\partial^2 E_y}{\partial y^2} + \frac{\partial}{\partial y}\left(\frac{1}{\varepsilon_r}\frac{\partial \varepsilon_r}{\partial y} E_y\right) + (k_0^2 \varepsilon_r - \beta^2) E_y = 0. \quad (4.130)$$

Comparing this equation with the E_x representation wave equation for the quasi-TE mode [Eq. (4.50)], we find that we can obtain the finite-difference expression for the E_y representation for the quasi-TM mode [Eq. (4.130)] by making the following replacements in Eqs. (4.100)–(4.105):

$$x \leftrightarrow y, \quad (4.131)$$

$$E_x \leftrightarrow E_y. \quad (4.132)$$

Thus, we get the following finite-difference expression for Eq. (4.130):

$$\alpha_w E_{p-1,q} + \alpha_e E_{p+1,q} + \alpha_n E_{p,q-1} + \alpha_s E_{p,q+1}$$
$$+ (\alpha_x + \alpha_y)E_{p,q} + \{k_0^2 \varepsilon_r(p,q) - \beta^2\}E_{p,q} = 0, \quad (4.133)$$

where

$$\alpha_w = \frac{2}{w(e+w)}, \quad (4.134)$$

$$\alpha_e = \frac{2}{e(e+w)}, \quad (4.135)$$

$$\alpha_n = \frac{2}{n(n+s)} \frac{2\varepsilon_r(p,q-1)}{\varepsilon_r(p,q) + \varepsilon_r(p,q-1)}, \quad (4.136)$$

$$\alpha_s = \frac{2}{s(n+s)} \frac{2\varepsilon_r(p,q+1)}{\varepsilon_r(p,q) + \varepsilon_r(p,q+1)}, \quad (4.137)$$

$$\alpha_x = -\frac{2}{ew} = -\alpha_e - \alpha_w, \quad (4.138)$$

$$\alpha_y = -\frac{2}{ns} - \frac{2}{n(n+s)} \frac{\varepsilon_r(p,q) - \varepsilon_r(p,q-1)}{\varepsilon_r(p,q) + \varepsilon_r(p,q-1)}$$
$$- \frac{2}{s(n+s)} \frac{\varepsilon_r(p,q) - \varepsilon_r(p,q+1)}{\varepsilon_r(p,q) + \varepsilon_r(p,q+1)}$$
$$= -\frac{4}{ns} + \alpha_n + \alpha_s. \quad (4.139)$$

B. H_x Representation The H_x representation wave equation for the quasi-TM mode [Eq. (4.43)] is

$$\frac{\partial^2 H_x}{\partial x^2} + \varepsilon_r \frac{\partial}{\partial y}\left(\frac{1}{\varepsilon_r}\frac{\partial H_x}{\partial y}\right) + (k_0^2 \varepsilon_r - \beta^2)H_x = 0. \quad (4.140)$$

Comparing this equation with the H_y representation for the quasi-TE mode [Eq. (4.106)], we find that the finite-difference expression for the H_x representation for the quasi-TM mode [Eq. (4.140)] can be obtained by making the following replacements in Eqs. (4.124)–(4.129):

$$x \leftrightarrow y, \quad (4.141)$$

$$H_y \leftrightarrow H_x. \quad (4.142)$$

Thus, we get the following finite-difference expression for the H_x representation:

$$\alpha_w H_{p-1,q} + \alpha_e H_{p+1,q} + \alpha_n H_{p,q-1} + \alpha_s H_{p,q+1}$$
$$+ (\alpha_x + \alpha_y)H_{p,q} + \{k_0^2 \varepsilon_r(p,q) - \beta^2\}H_{p,q} = 0, \quad (4.143)$$

where

$$\alpha_w = \frac{2}{w(e+w)}, \quad (4.144)$$

$$\alpha_e = \frac{2}{e(w+e)}, \quad (4.145)$$

$$\alpha_n = \frac{2}{n(n+s)} \frac{2\varepsilon_r(p,q)}{\varepsilon_r(p,q) + \varepsilon_r(p,q-1)}, \quad (4.146)$$

$$\alpha_s = \frac{2}{s(n+s)} \frac{2\varepsilon_r(p,q)}{\varepsilon_r(p,q) + \varepsilon_r(p,q+1)}, \quad (4.147)$$

$$\alpha_x = -\frac{2}{ew} = -\alpha_e - \alpha_w, \quad (4.148)$$

$$\alpha_y = -\frac{2}{n(n+s)} \frac{2\varepsilon_r(p,q)}{\varepsilon_r(p,q) + \varepsilon_r(p,q-1)}$$
$$-\frac{2}{s(n+s)} \frac{2\varepsilon_r(p,q)}{\varepsilon_r(p,q) + \varepsilon_r(p,q+1)}$$
$$= -\alpha_n - \alpha_s. \quad (4.149)$$

4.3.3 Scalar Mode

The scalar wave equation [Eq. (4.49)] is

$$\frac{\partial^2 \phi}{\partial x^2} + \frac{\partial^2 \phi}{\partial y^2} + (k_0^2 \varepsilon_r - \beta^2)\phi = 0. \quad (4.150)$$

Since the derivatives of relative permittivity with respect to x or y in the semivectorial wave equations are set to zero for the scalar analysis, we can easily get the finite-difference expression for the scalar finite-difference method (SC-FDM):

$$\alpha_w \phi_{p-1,q} + \alpha_e \phi_{p+1,q} + \alpha_n \phi_{p,q-1} + \alpha_s \phi_{p,q+1}$$
$$+ (\alpha_x + \alpha_y)\phi_{p,q} + \{k_0^2 \varepsilon_r(p,q) - \beta^2\}\phi_{p,q} = 0, \quad (4.151)$$

where

$$\alpha_w = \frac{2}{w(e+w)}, \quad (4.152)$$

$$\alpha_e = \frac{2}{e(w+e)}, \quad (4.153)$$

$$\alpha_n = \frac{2}{n(n+s)}, \quad (4.154)$$

$$\alpha_s = \frac{2}{s(n+s)}, \quad (4.155)$$

$$\alpha_x = -\frac{2}{ew} = -\alpha_e - \alpha_w, \quad (4.156)$$

$$\alpha_y = -\frac{2}{ns} = -\alpha_n - \alpha_s. \quad (4.157)$$

4.4 PROGRAMMING

Now, we look at how an eigenvalue matrix equation can be solved using the FDM. The procedure is almost the same as that for the solution using the FEM, described in detail in Chapter 3, so only important differences are dealt with here.

The finite-difference expression for the semivectorial wave equation was obtained by approximating the derivatives with the difference expressions:

$$\alpha_w \phi_{p-1,q} + \alpha_e \phi_{p+1,q} + \alpha_n \phi_{p,q-1} + \alpha_s \phi_{p,q+1}$$
$$+ \{(\alpha_x + \alpha_y) + k_0^2 \varepsilon_r(p,q)\}\phi_{p,q} - \beta^2 \phi_{p,q} = 0, \quad (4.158)$$

where (p,q) corresponds to coordinates (x,y) and where $\phi_{p,q}$ corresponds, for the quasi-TE mode, to the field component E_x or H_y and corresponds, for the quasi-TM mode, to the field component E_y or H_x.

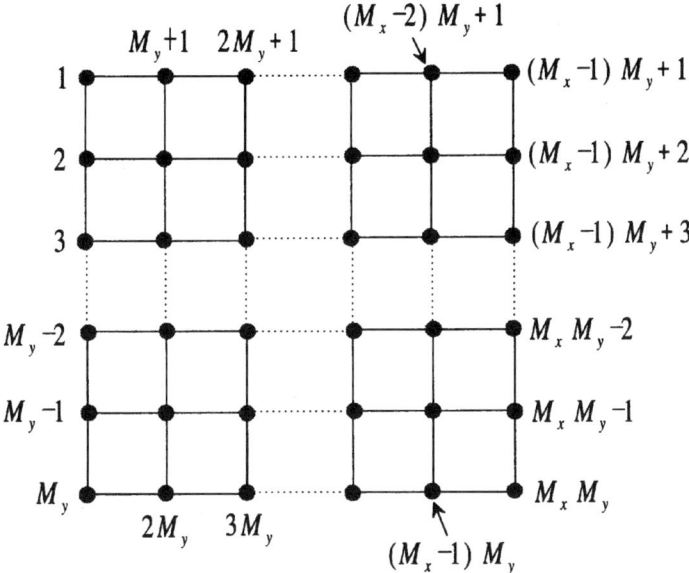

FIGURE 4.4. Meshes for the finite-difference method.

Figure 4.4 shows a mesh model in a finite-difference scheme in which the whole analysis area is divided into a number of meshes and the nodes are numbered from top to bottom and from left to right. Calculating Eq. (4.158) for each node in an $M_x \times M_y$ matrix, we obtain the following eigenvalue matrix equation:

$$[A]\{\phi\} = \beta^2 \{\phi\}. \tag{4.159}$$

Here, β^2 is an eigenvalue and $\{\phi\}$ is an eigenvector expressed as

$$\{\phi\} = (\phi_1 \quad \phi_2 \quad \phi_3 \quad \cdots \quad \phi_M)^T, \tag{4.160}$$

where M is $M_x \times M_y$. It should be noted that, if the variable transformations $\bar{x} = xk_0$ and $\bar{y} = yk_0$ mentioned in Section 4.4 are used, the eigenvalue in Eq. (4.159) is n_{eff}^2, where $n_{\text{eff}}(\beta/k_0)$ is the effective index. Figure 4.5 shows the global matrix $[A]$ corresponding to the meshes shown in Fig. 4.4.

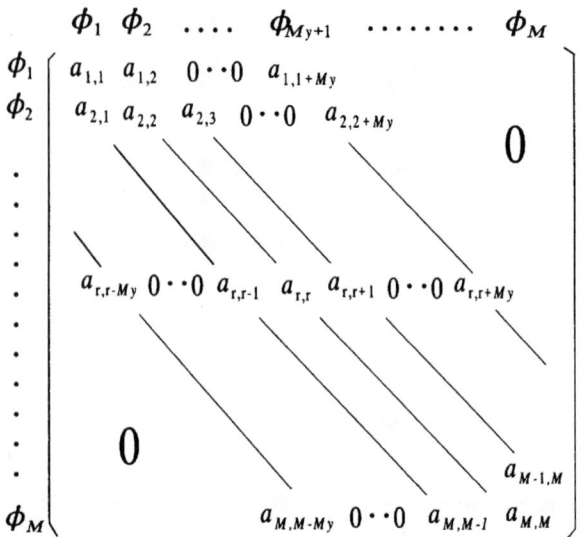

FIGURE 4.5. Form of global matrix [A]. Here, $M = M_x M_y$.

We can obtain the propagation constant and field distribution by solving the eigenvalue matrix equation (4.159). In the SV-FDMs, the global matrix [A] is a nonsymmetric sparse matrix. In the SC-FDM, on the other hand, the global matrix is symmetric, so only half of it has to be calculated. Taking these facts into account and considering the interaction between nodes shown in Figs. 4.4 and 4.5, we can easily understand that the bandwidth of the global matrix is $2M_y + 1$ in the SV-FDM and is $M_y + 1$ in the SC-FDM. The latter bandwidth is the same as that of the global matrix in the first-order SC-FEM.

In the actual programming, the node number r is used instead of (p, q). When nodes are numbered from top to bottom and from left to right as shown in Fig. 4.4, the node number r for (p, q) can be designated as follows:

$$r = (p-1)M_y + q, \qquad (4.161)$$

where $1 \leq p \leq M_x$ and $1 \leq q \leq M_y$. The rth row in the eigenvalue matrix equation can be expressed as

$$a_{r,r-M_y}\phi_{r-M_y} + a_{r,r-1}\phi_{r-1} + a_{r,r}\phi_r + a_{r,r+1}\phi_{r+1}$$
$$+ a_{r,r+M_y}\phi_{r+M_y} - \beta^2 \phi_r = 0, \qquad (4.162)$$

where $a_{i,j}$ is an element of the global matrix $[A]$. The coefficients of Eq. (4.158) and Eq. (4.162) correspond as follows:

$$\alpha_w \leftrightarrow a_{r,r-M_y}, \tag{4.163}$$

$$\alpha_e \leftrightarrow a_{r,r+M_y}, \tag{4.164}$$

$$\alpha_n \leftrightarrow a_{r,r-1}, \tag{4.165}$$

$$\alpha_s \leftrightarrow a_{r,r+1}, \tag{4.166}$$

$$\alpha_x + \alpha_y + k_0^2 \varepsilon_r(p,q) \leftrightarrow a_{r,r}. \tag{4.167}$$

The eigenvalue matrix equation is formed in basically the same way as that used to form the eigenvalue matrix equation for the FEM discussed in Section 3.6. There is, however, one important difference. In the FEM, the whole analysis area is divided into a number of elements, and the variational principle or the Galerkin method is applied to each element. Therefore, as shown in Eqs. (3.293) and (3.294), when the FEM is used, the formation of the global matrix requires a summation over all the elements. In the FDM, on the other hand, the derivatives in the wave equations are approximated with finite differences. The FDM thus does not require a summation in order to form the global matrix.

4.5 BOUNDARY CONDITIONS

In the actual programming, we have to impose boundary conditions on the nodes on the edge of the analysis window. In other words, it is necessary that the effect of nodes outside the analysis window be taken into account at the edge of the window. Here, the Dirichlet, Neumann, and analytical boundary conditions will be discussed.

DIRICHLET CONDITION   A wave function outside the analysis window is set to zero. Thus,

$$\phi = 0. \tag{4.168}$$

NEUMANN CONDITION   The normal derivative of a wave function at the edge of the analysis window is set to zero. Thus,

$$\frac{\partial \phi}{\partial n} = 0. \tag{4.169}$$

In other words, it is assumed that the value of a hypothetical wave function outside the analysis window is equal to that of an actual wave function at the edge of the analysis window.

ANALYTICAL BOUNDARY CONDITION When we assume that the wave number in a vacuum, the effective index, the discretization width at the edge of the analysis window, and the relative permittivity are respectively k_0, n_{eff}, Δ and $\varepsilon_r(p,q)$, the analytical wave function outside the analysis window to be connected with a wave function at the edge of the analysis window is assumed to decay exponentially with the decay constant $-k_0\sqrt{|n_{\text{eff}}^2 - \varepsilon_r(p,q)|}$:

$$\exp\left(-k_0\sqrt{|n_{\text{eff}}^2 - \varepsilon_r(p,q)|} \cdot \Delta\right). \tag{4.170}$$

WRITING BOUNDARY CONDITIONS INTO PROGRAMS Let us consider here how the boundary conditions can be written into a computer program. Although Eq. (4.162) has to be used for programming, for simplicity we will instead consider Eq. (4.158), written here as

$$\alpha_w \phi_{p-1,q} + \alpha_e \phi_{p+1,q} + \alpha_n \phi_{p,q-1} + \alpha_s \phi_{p,q+1}$$
$$+ \{(\alpha_x + \alpha_y) + k_0^2 \varepsilon_r(p,q)\} \phi_{p,q} - \beta^2 \phi_{p,q} = 0. \tag{4.171}$$

a. Left-Hand Boundary ($p = 1$ and except at corners) Consider the left-hand boundary shown as 1 in Fig. 4.6. Here, the discretization width is Δx. When we assume that (p,q) is a node on the boundary, the hypothetical node outside the analysis window is $(p-1,q)$ and we get

$$\phi_{p-1,q} = \gamma_L \phi_{p,q}. \tag{4.172}$$

For the Dirichlet, Neumann, and analytical conditions, the coefficient γ_L is expressed as

$$\gamma_L = \begin{cases} 0 & \text{(Dirichlet)}, \\ 1 & \text{(Neumann)}, \\ \exp(-k_0\sqrt{|n_{\text{eff}}^2 - \varepsilon_r(p,q)|} \cdot \Delta x) & \text{(analytical)}. \end{cases} \tag{4.173}$$

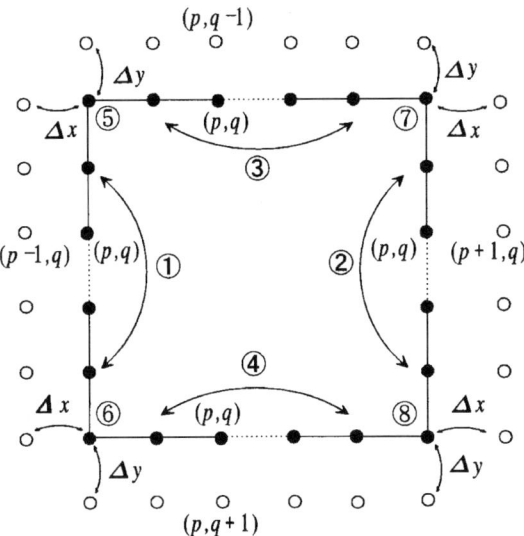

FIGURE 4.6. Nodes on boundaries and hypothetical nodes outside boundaries.

Substituting Eqs. (4.172) and (4.173) into Eq. (4.171), we get

$$\alpha_e \phi_{p+1,q} + \alpha_n \phi_{p,q-1} + \alpha_s \phi_{p,q+1} \\ + \{\alpha_w \gamma_L + (\alpha_x + \alpha_y) + k_0^2 \varepsilon_r(p,q)\} \phi_{p,q} - \beta^2 \phi_{p,q} = 0. \quad (4.174)$$

b. Right-Hand Boundary ($p = M_x$ and except at corners) Consider the right-hand boundary shown as 2 in Fig. 4.6. Here, the discretization width is again Δx. When we assume that (p, q) is a node on the boundary, the hypothetical node outside the analysis window is $(p+1, q)$ and we get

$$\phi_{p+1,q} = \gamma_R \phi_{p,q}. \quad (4.175)$$

For the Dirichlet, Neumann, and analytical conditions, the coefficient γ_R is expressed as

$$\gamma_R = \begin{cases} 0 & \text{(Dirichlet)}, \\ 1 & \text{(Neumann)}, \\ \exp(-k_0 \sqrt{|n_{\text{eff}}^2 - \varepsilon_r(p,q)|} \cdot \Delta x) & \text{(analytical)}. \end{cases} \quad (4.176)$$

Substituting Eqs. (4.175) and Eq. (4.176) into Eq. (4.171), we get

$$\alpha_w \phi_{p-1,q} + \alpha_n \phi_{p,q-1} + \alpha_s \phi_{p,q+1}$$
$$+ \{\alpha_e \gamma_R + (\alpha_x + \alpha_y) + k_0^2 \varepsilon_r(p,q)\} \phi_{p,q} - \beta^2 \phi_{p,q} = 0. \quad (4.177)$$

c. Top Boundary ($q = 1$ and except at corners) Consider the top boundary shown as 3 in Fig. 4.6. Here, the discretization width is Δy. When we assume that (p, q) is a node on the boundary, the hypothetical node outside the analysis window is $(p, q-1)$ and we get

$$\phi_{p,q-1} = \gamma_U \phi_{p,q}. \quad (4.178)$$

For the Dirichlet, Neumann, and analytical conditions, the coefficient γ_U is expressed as

$$\gamma_U = \begin{cases} 0 & \text{(Dirichlet)}, \\ 1 & \text{(Neumann)}, \\ \exp(-k_0 \sqrt{|n_{\text{eff}}^2 - \varepsilon_r(p,q)|} \cdot \Delta y) & \text{(analytical)}. \end{cases} \quad (4.179)$$

Substituting Eqs. (4.178) and (4.179) into Eq. (4.171), we get

$$\alpha_w \phi_{p-1,q} + \alpha_e \phi_{p+1,q} + \alpha_s \phi_{p,q+1}$$
$$+ \{\alpha_n \gamma_U + (\alpha_x + \alpha_y) + k_0^2 \varepsilon_r(p,q)\} \phi_{p,q} - \beta^2 \phi_{p,q} = 0. \quad (4.180)$$

d. Bottom Boundary ($q = M_y$ and except at corners) Consider the bottom boundary shown as 4 in Fig. 4.6. Here, the discretization width is again Δy. When we assume that (p, q) is a node on the boundary, the hypothetical node outside the analysis window is $(p, q+1)$ and we get

$$\phi_{p,q+1} = \gamma_D \phi_{p,q}. \quad (4.181)$$

For the Dirichlet, Neumann, and analytical conditions, the coefficient γ_D is expressed as

$$\gamma_D = \begin{cases} 0 & \text{(Dirichlet)}, \\ 1 & \text{(Neumann)}, \\ \exp(-k_0 \sqrt{|n_{\text{eff}}^2 - \varepsilon_r(p,q)|} \cdot \Delta y) & \text{(analytical)}. \end{cases} \quad (4.182)$$

Substituting Eqs. (4.181) and (4.182) into Eq. (4.171), we get

$$\alpha_w \phi_{p-1,q} + \alpha_e \phi_{p+1,q} + \alpha_n \phi_{p,q-1}$$
$$+ \{\alpha_s \gamma_D + (\alpha_x + \alpha_y) + k_0^2 \varepsilon_r(p,q)\} \phi_{p,q} - \beta^2 \phi_{p,q} = 0. \quad (4.183)$$

e. Left-Top Corner ($p = q = 1$) Consider the left-top corner shown as 5 in Fig. 4.6. Here, the discretization widths are Δx and Δy. When we assume that (p,q) is the node at the corner, the hypothetical nodes outside the analysis window are $(p-1,q)$ to the left and $(p,q-1)$ to the top. We thus get

$$\phi_{p-1,q} = \gamma_L \phi_{p,q}, \quad (4.184)$$

$$\phi_{p,q-1} = \gamma_U \phi_{p,q}. \quad (4.185)$$

For the Dirichlet, Neumann, and analytical, conditions, the coefficients γ_L and γ_U are expressed as

$$\gamma_L = \begin{cases} 0 & \text{(Dirichlet)}, \\ 1 & \text{(Neumann)}, \\ \exp(-k_0\sqrt{|n_{\text{eff}}^2 - \varepsilon_r(p,q)|} \cdot \Delta x) & \text{(analytical)}, \end{cases} \quad (4.186)$$

$$\gamma_U = \begin{cases} 0 & \text{(Dirichlet)}, \\ 1 & \text{(Neumann)}, \\ \exp(-k_0\sqrt{|n_{\text{eff}}^2 - \varepsilon_r(p,q)|} \cdot \Delta y) & \text{(analytical)}. \end{cases} \quad (4.187)$$

Substituting Eqs. (4.184) to (4.187) into Eq. (4.171), we get

$$\alpha_e \phi_{p+1,q} + \alpha_s \phi_{p,q+1}$$
$$+ \{\alpha_w \gamma_L + \alpha_n \gamma_U + (\alpha_x + \alpha_y) + k_0^2 \varepsilon_r(p,q)\} \phi_{p,q} - \beta^2 \phi_{p,q} = 0. \quad (4.188)$$

f. Left-Bottom Corner ($p = 1, q = M_y$) Consider the left- bottom corner shown as 6 in Fig. 4.6. Here, the discretization widths are again Δx and Δy. When we assume that (p,q) is the node at the corner, the hypothetical nodes outside the analysis window are $(p-1,q)$ to the left and $(p,q+1)$ to the bottom. We thus get

$$\phi_{p-1,q} = \gamma_L \phi_{p,q}, \quad (4.189)$$

$$\phi_{p,q+1} = \gamma_D \phi_{p,q}. \quad (4.190)$$

For the Dirichlet, Neumann, and analytical conditions, the coefficients γ_L and γ_D are expressed as

$$\gamma_L = \begin{cases} 0 & \text{(Dirichlet)}, \\ 1 & \text{(Neumann)}, \\ \exp(-k_0\sqrt{|n_{\text{eff}}^2 - \varepsilon_r(p,q)|} \cdot \Delta x) & \text{(analytical)}, \end{cases} \quad (4.191)$$

$$\gamma_D = \begin{cases} 0 & \text{(Dirichlet)}, \\ 1 & \text{(Neumann)}, \\ \exp(-k_0\sqrt{|n_{\text{eff}}^2 - \varepsilon_r(p,q)|} \cdot \Delta y) & \text{(analytical)}. \end{cases} \quad (4.192)$$

Substituting Eqs. (4.189) to (4.192) into Eq. (4.171), we get

$$\alpha_e \phi_{p+1,q} + \alpha_n \phi_{p,q-1} \\ + \{\alpha_w \gamma_L + \alpha_s \gamma_D + (\alpha_x + \alpha_y) + k_0^2 \varepsilon_r(p,q)\}\phi_{p,q} - \beta^2 \phi_{p,q} = 0. \quad (4.193)$$

g. Right-Top Corner ($p = M_x, q = 1$) Consider the right-top corner shown as 7 in Fig. 4.6. Here, the discretization widths are again Δx and Δy. When we assume that (p, q) is the node at the corner, the hypothetical nodes outside the analysis window are $(p+1, q)$ to the right and $(p, q-1)$ to the top. We thus get

$$\phi_{p+1,q} = \gamma_R \phi_{p,q}, \quad (4.194)$$

$$\phi_{p,q-1} = \gamma_U \phi_{p,q}. \quad (4.195)$$

For the Dirichlet, Neumann, and analytical conditions, the coefficients γ_R and γ_U are expressed as

$$\gamma_R = \begin{cases} 0 & \text{(Dirichlet)}, \\ 1 & \text{(Neumann)}, \\ \exp(-k_0\sqrt{|n_{\text{eff}}^2 - \varepsilon_r(p,q)|} \cdot \Delta x) & \text{(analytical)}, \end{cases} \quad (4.196)$$

$$\gamma_U = \begin{cases} 0 & \text{(Dirichlet)}, \\ 1 & \text{(Neumann)}, \\ \exp(-k_0\sqrt{|n_{\text{eff}}^2 - \varepsilon_r(p,q)|} \cdot \Delta y) & \text{(analytical)}. \end{cases} \quad (4.197)$$

Substituting Eqs. (4.194)–(4.197) into Eq. (4.171), we get

$$\alpha_w \phi_{p-1,q} + \alpha_s \phi_{p,q+1}$$
$$+ \{\alpha_e \gamma_R + \alpha_n \gamma_U + (\alpha_x + \alpha_y) + k_0^2 \varepsilon_r(p,q)\}\phi_{p,q} - \beta^2 \phi_{p,q} = 0. \quad (4.198)$$

h. Right-Bottom Corner ($p = M_x, q = M_y$) Consider the right-bottom corner shown as 8 in Fig. 4.6. Here, the discretization widths are again Δx and Δy. When we assume that (p,q) is the node at the corner, the hypothetical nodes outside the analysis window are $(p+1, q)$ to the right and $(p, q+1)$ to the bottom. We thus get

$$\phi_{p+1,q} = \gamma_R \phi_{p,q}, \quad (4.199)$$

$$\phi_{p,q+1} = \gamma_D \phi_{p,q}. \quad (4.200)$$

For the Dirichlet, Neumann, and analytical conditions, the coefficients γ_R and γ_D are expressed as

$$\gamma_R = \begin{cases} 0 & \text{(Dirichlet)}, \\ 1 & \text{(Neumann)}, \\ \exp(-k_0\sqrt{|n_{\text{eff}}^2 - \varepsilon_r(p,q)|} \cdot \Delta x) & \text{(analytical)}, \end{cases} \quad (4.201)$$

$$\gamma_D = \begin{cases} 0 & \text{(Dirichlet)}, \\ 1 & \text{(Neumann)}, \\ \exp(-k_0\sqrt{|n_{\text{eff}}^2 - \varepsilon_r(p,q)|} \cdot \Delta y) & \text{(analytical)}. \end{cases} \quad (4.202)$$

Substituting Eqs. (4.199)–(4.202) into Eq. (4.171), we get

$$\alpha_w \phi_{p-1,q} + \alpha_n \phi_{p,q-1}$$
$$+ \{\alpha_e \gamma_R + \alpha_s \gamma_D + (\alpha_x + \alpha_y) + k_0^2 \varepsilon_r(p,q)\}\phi_{p,q} - \beta^2 \phi_{p,q} = 0. \quad (4.203)$$

In a way similar to that described in Section 3.7 for the FEM, we can calculate the even or the odd mode by assuming Neumann or Dirichlet conditions at the symmetry plane. This increases the numerical efficiency in terms of central processing unit (CPU) time and computer memory.

4.6 NUMERICAL EXAMPLE

The finite-difference expressions in this chapter are used in CAD software currently available on the market. Readers can also develop software by themselves. This section briefly discusses a calculation model and results calculated using SV-FDM software. Figure 4.7 shows a calculation model that has a 0.4-μm^2 core. The refractive indexes for a wavelength of 1.55 μm are 3.5 for the core and 3.1693 for the cladding. The non-equidistant discretization scheme was used in this example. The number of nodes M_x in the horizontal direction and the number of nodes M_y in the vertical direction were both 96, and the minimum and maximum discretization widths in both directions were respectively 0.025 and 0.05 μm.

The calculated effective index n_{eff} for both the quasi-TE and quasi-TM modes was 3.2172, and Fig. 4.8 is a three-dimensional plot of the electric field component E_x calculated for the quasi-TE mode. It should be noted that since the normal component of the electric flux density $D_x = \varepsilon_r E_x$ is continuous at the interface between two media, as shown in Eq. (1.55), the normal component of the electric field E_x is not continuous at the interface. In Fig. 4.8, we can clearly see the discontinuities of E_x.

Since, as pointed out in Sections 4.1 and 4.3, the finite-difference expressions have the errors of the order of h for nonequidistant discretization and of the order of h^2 for equidistant discretization, equidistant discretization is preferable. But because extremely fine meshes are necessary for dealing with interfaces at which the refractive index changes abruptly, as in this calculation model, nonequidistant discretization must

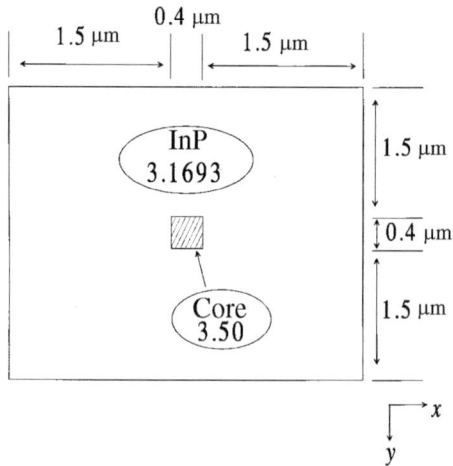

FIGURE 4.7. Calculation model.

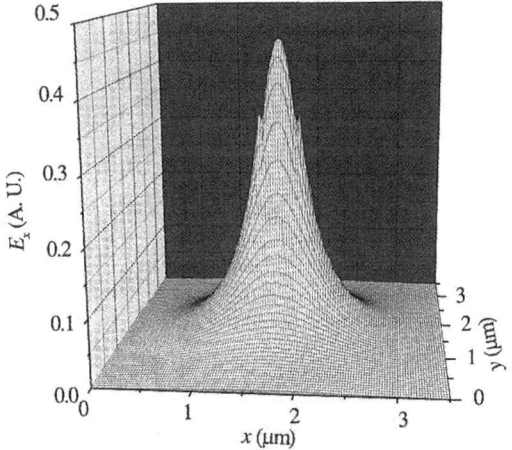

FIGURE 4.8. The x-directed electric field component E_x.

be used in order to reduce the amounts of computer memory and CPU time required. The scalar FEM did not give the degenerated results for the square core due to inaccurate boundary conditions, as discussed in Chapter 3. The semivectorial FDMs, on the other hand, can overcome this difficulty for a small-aspect core structure.

PROBLEMS

1. Show that when the second derivative is approximated by Eq. (4.9),

$$f^{(2)}(0) = \frac{2}{h_1 h_2} \frac{h_2 f_1 - (h_1 + h_2) f_3 + h_1 f_2}{(h_1 + h_2)}, \quad \text{(P4.1)}$$

the error is $O(h^2)$ when $h_1 = h_2$ (equidistant discretization) and is $O(h)$ when $h_1 \neq h_2$ (nonequidistant discretization).

ANSWER

Multiplying Eq. (4.4) by h_2 and Eq. (4.5) by h_1 and adding them, we get

$$f^{(2)}(0) = \frac{2}{h_1 h_2} \frac{h_2 f_1 - (h_1 + h_2) f_3 + h_1 f_2}{(h_1 + h_2)} - \frac{1}{3}(h_2 - h_1) f^{(3)}(0) + O(h^2), \quad \text{(P4.2)}$$

from which the answer can easily be obtained.

2. Derive the finite-difference expression (4.121) without using the Taylor series expansion.

ANSWER

Since the hypothetical difference center is $(p+\frac{1}{2},q)$ for the first term within the brackets on the right-hand side of Eq. (4.110), we can derive the equations

$$\varepsilon_r(p+\tfrac{1}{2},q) = \tfrac{1}{2}\{\varepsilon_r(p+1,q) + \varepsilon_r(p,q)\} \tag{P4.3}$$

and

$$\left.\frac{\partial H_y}{\partial x}\right|_{p+1/2,q} = \frac{1}{e}\{H_y(p+1,q) - H_y(p,q)\}, \tag{P4.4}$$

from which we can immediately derive

$$\frac{1}{\varepsilon_r}\left.\frac{\partial H_y}{\partial x}\right|_{p+1/2,q} = \frac{1}{e}\frac{2}{\varepsilon_r(p+1,q)+\varepsilon_r(p,q)}(H_{p+1,q} - H_{p,q}). \tag{P4.5}$$

And since the hypothetical difference center is $(p-\frac{1}{2},q)$ for the second term within the brackets, we can derive the equations

$$\varepsilon_r(p-\tfrac{1}{2},q) = \tfrac{1}{2}\{\varepsilon_r(p,q) - \varepsilon_r(p-1,q)\} \tag{P4.6}$$

and

$$\left.\frac{\partial H_y}{\partial x}\right|_{p-1/2,q} = \frac{1}{w}\{H_y(p,q) - H_y(p-1,q)\}, \tag{P4.7}$$

from which we can immediately derive

$$\frac{1}{\varepsilon_r}\left.\frac{\partial H_y}{\partial x}\right|_{p-1/2,q} = \frac{1}{w}\frac{2}{\varepsilon_r(p,q)+\varepsilon_r(p-1,q)}(H_{p,q} - H_{p-1,q}). \tag{P4.8}$$

Substituting Eqs. (P4.5) and (P4.8) into Eq. (4.110) and using some mathematical manipulations, we can easily derive Eq. (4.121).

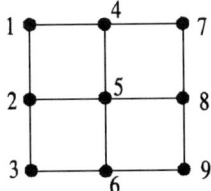

FIGURE P4.1. Simple example of the finite-difference method.

3. A simple calculation model for the semivectorial FDM is shown in Fig. P4.1. Show the form of the global matrix $[A]$ of Eq. (4.159).

ANSWER

The global matrix $[A]$ is shown in Fig. P4.2. It should be noted that the interactions between nodes 3 and 4, which correspond to $a_{3,4}$ and $a_{4,3}$, and the interactions between nodes 6 and 7, which correspond to $a_{6,7}$ and $a_{7,6}$, have to be set to zero. Since the number of nodes in the y direction is 3, the bandwidth is 7. The reason for this is explained in Section 4.4.

4. Calculate effective indexes n_{eff} for the quasi-TE and quasi-TM modes of the strip-loaded optical waveguide shown in Fig. P4.3. Assume that the refractive indexes for a wavelength of 1.55 μm are 3.3884 for the InGaAsP of the 1.3-μm band-gap wavelength (1.3Q) core and 3.1693 for the InP cladding.

$$[A] = \begin{bmatrix} a_{1,1} & a_{1,2} & 0 & a_{1,4} & & & & & \\ a_{2,1} & a_{2,2} & a_{2,3} & 0 & a_{2,5} & & 0 & & \\ 0 & a_{3,2} & a_{3,3} & 0 & 0 & a_{3,6} & & & \\ a_{4,1} & 0 & 0 & a_{4,4} & a_{4,5} & 0 & a_{4,7} & & \\ & a_{5,2} & 0 & a_{5,4} & a_{5,5} & a_{5,6} & 0 & a_{5,8} & \\ & & a_{6,3} & 0 & a_{6,5} & a_{6,6} & 0 & 0 & a_{6,9} \\ & & & a_{7,4} & 0 & 0 & a_{7,7} & a_{7,8} & 0 \\ & 0 & & & a_{8,5} & 0 & a_{8,7} & a_{8,8} & a_{8,9} \\ & & & & & a_{9,6} & 0 & a_{9,8} & a_{9,9} \end{bmatrix}$$

FIGURE P4.2. Form of global matrix $[A]$.

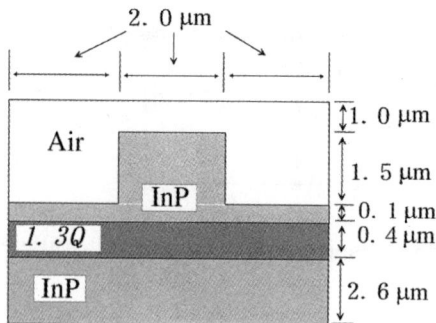

FIGURE P4.3. Calculation example of a strip-loaded optical waveguide.

ANSWER

The results you get will slightly depend on the CAD software used to calculate them, but the effective index for the quasi-TE mode is 3.2576 and that for the quasi-TM mode is 3.2471.

REFERENCES

[1] M. Stern, "Semivectorial polarized finite difference method optical waveguides with arbitrary index profiles," *IEE Proc. J.*, vol. 135, pp. 56–63, 1988.
[2] M. Stern, "Semivectorial polarized H field solutions for dielectric waveguides with arbitrary index profiles," *IEE Proc. J.*, vol. 135, pp. 333–338, 1988.
[3] K. Kawano, T. Kitoh, M. Kohtoku, T. Takeshita, and Y. Hasumi, "3-D semivectorial analysis to calculate facet reflectivities of semiconductor optical waveguides based on the bi-directional method of line BPM (MoL- BPM)," *IEEE Photon. Technol. Lett.*, vol. 10, pp. 108–110, 1998.
[4] B. Gerdes, B. Lunitz, D. Benish, and R. Pregla, "Analysis of slab waveguide discontinuities including radiation and absorption effects," *Electron. Lett.*, vol. 28, pp. 1013–1015, 1992.

CHAPTER 5

BEAM PROPAGATION METHODS

The analysis methods discussed in the preceding chapters assumed the structures of the optical waveguides to be uniform in the propagation direction. A lot of the waveguides used in actual optical waveguide devices, however, have nonuniform structures such as bends, tapers, and crosses in the propagation direction. In this chapter, we discuss the beam propagation methods (BPMs) that have been developed for the analysis of such nonuniform structures.

Various kinds of BPMs, such as the fast Fourier transform (FFT-BPM) [1–4], the finite difference (FD-BPM) [4–11], and the finite element (FE-BPM) [12], have been developed. For the derivatives with respect to the coordinates in the lateral directions, they respectively make use of the fast Fourier transform (FFT), the finite-difference (FD) approximation, and the finite-element (FE) approximation. The FFT-BPM and the FD-BPM will be discussed here. Beam propagation CAD software is widely available on the market.

5.1 FAST FOURIER TRANSFORM BEAM PROPAGATION METHOD

The FFT-BPM [1] had been widely applied to design optical waveguides until the FD-BPM [4] was developed. The FFT-BPM has the following disadvantages due to the nature of the FFT: (1) it requires a long computation time, (2) the discretization widths in the lateral directions must be uniform, (3) the simple transparent boundary condition cannot be used at the analysis boundaries, (4) very small discretization widths cannot be used in the lateral directions, (5) the polarization cannot be treated, (6) it is inadequate for large-index-difference optical waveguides, (7) the number of sampling points must be a power of 2, and (8) the propagation step has to be small. But it is investigated here because the FFT-BPM is historically important and the line of thinking it exemplifies is very interesting and useful.

5.1.1 Wave Equation

The scalar Helmholtz equation is expressed as

$$\nabla^2 \psi(x, y, z) + k_0^2 n^2(x, y, z)\psi(x, y, z) = 0, \qquad (5.1)$$

where ∇^2 is the Laplacian

$$\nabla^2 = \frac{\partial^2}{\partial x^2} + \frac{\partial^2}{\partial y^2} + \frac{\partial^2}{\partial z^2} \qquad (5.2)$$

and k_0 is the wave number in a vacuum.

Here, the slowly varying envelope approximation (SVEA) is used to approximate the wave function $\psi(x, y, z)$ of the light propagating in the $+z$ direction. In this approximation, $\psi(x, y, z)$ is separated into the slowly varying envelope function $\phi(x, y, z)$ and the very fast oscillatory phase term $\exp(-j\beta z)$ as follows:

$$\psi(x, y, z) = \phi(x, y, z) \exp(-j\beta z). \qquad (5.3)$$

Here,

$$\beta = n_{\text{eff}} k_0, \qquad (5.4)$$

where n_{eff} is the reference index, for which the refractive index of the substrate or cladding is usually used.

Substituting the second derivative of the wave function $\psi(x, y, z)$ with respect to z,

$$\frac{\partial^2 \psi}{\partial z^2} = \frac{\partial^2 \phi}{\partial z^2} \exp(-j\beta z) - 2j\beta \frac{\partial \phi}{\partial z} \exp(-j\beta z) - \beta^2 \phi \exp(-j\beta z), \qquad (5.5)$$

into Eq. (5.1) and dividing both sides by the exponential term $\exp(-j\beta z)$, we get

$$\frac{\partial^2 \phi}{\partial z^2} - 2j\beta \frac{\partial \phi}{\partial z} + \nabla_\perp^2 \phi + (k_0^2 n^2 - \beta^2)\phi = 0, \tag{5.6}$$

where $\nabla_\perp^2$ is a Laplacian in the lateral directions (i.e., the x and y directions) and is expressed as

$$\nabla_\perp^2 = \frac{\partial^2}{\partial x^2} + \frac{\partial^2}{\partial y^2}. \tag{5.7}$$

Or, using the relation

$$k_0^2 n^2 - \beta^2 = k_0^2 (n^2 - n_{\text{eff}}^2), \tag{5.8}$$

we get

$$2j\beta \frac{\partial \phi}{\partial z} - \frac{\partial^2 \phi}{\partial z^2} = \nabla_\perp^2 \phi + k_0^2 (n^2 - n_{\text{eff}}^2)\phi. \tag{5.9}$$

Since the second derivative of the wave function ϕ with respect to z is not neglected, Eq. (5.9) is a wide-angle formulation. On the other hand, when the second derivative is neglected, that is, when we assume

$$\frac{\partial^2 \phi}{\partial z^2} = 0, \tag{5.10}$$

the wave equation (5.9) is reduced to

$$2j\beta \frac{\partial \phi}{\partial z} = \nabla_\perp^2 \phi + k_0^2 (n^2 - n_{\text{eff}}^2)\phi. \tag{5.11}$$

The assumption that the second derivative of the wave function ϕ with respect to z can be neglected is called the Fresnel approximation or the para-axial approximation. The equality of the Fresnel approximation to the para-axial approximation will be discussed in Section 5.1.5.

5.1.2 Fresnel Approximation

Let us try to solve the wave equation (5.11), which is based on the Fresnel approximation. First, we separate the variables of the wave function ϕ of the Fresnel wave equation into the propagation direction and the lateral directions:

$$\phi(x, y, z) = A(x, y)\exp(\gamma z). \tag{5.12}$$

Substituting Eq. (5.12) into Eq. (5.11) and dividing both sides by ϕ, we get

$$2j\beta\gamma = \nabla_\perp^2 + k_0^2(n^2 - n_{\text{eff}}^2),$$

and therefore

$$\gamma = -\frac{j}{2\beta}\{\nabla_\perp^2 + k_0^2(n^2 - n_{\text{eff}}^2)\}. \tag{5.13}$$

Substituting γ in Eq. (5.13) into Eq (5.12), we get

$$\phi(x, y, z) = A(x, y)\exp\left(-\frac{j}{2\beta}\{\nabla_\perp^2 + k_0^2(n^2 - n_{\text{eff}}^2)\}z\right). \tag{5.14}$$

Thus, the wave function $\phi(x, y, z + \Delta z)$, which advances further than $\phi(x, y, z)$ by Δz in the propagation direction, can be written as

$$\phi(x, y, z + \Delta z) = \exp\left(-\frac{j}{2\beta}\{\nabla_\perp^2 + k_0^2(n^2 - n_{\text{eff}}^2)\}\Delta z\right)\phi(x, y, z). \tag{5.15}$$

We separate the exponential term into the following two terms:

$$\phi(x, y, z + \Delta z) = \exp\left(-j\frac{\Delta z}{2\beta}\nabla_\perp^2\right)\exp(-j\chi)\phi(x, y, z), \tag{5.16}$$

where

$$\chi = \frac{1}{2\beta} k_0^2 (n^2 - n_{\text{eff}}^2) \Delta z$$

$$= \frac{k_0^2}{2k_0 n_{\text{eff}}} \{(n_{\text{eff}} + \Delta n)^2 - n_{\text{eff}}^2\} \Delta z$$

$$= k_0 \Delta n \Delta z. \tag{5.17}$$

Here, we used the relation $n = n_{\text{eff}} + \Delta n$ and, assuming that Δn is sufficiently small, neglected $(\Delta n)^2$.

Since $\nabla_\perp^2$ in Eq. (5.16) is a *derivative operator*, the following relation holds for the general function f:

$$\nabla_\perp^2 \Delta n f - (\Delta n \nabla_\perp^2) f = (\nabla_\perp^2 \Delta n) f + \Delta n (\nabla_\perp^2 f) - (\Delta n \nabla_\perp^2) f$$

$$= (\nabla_\perp^2 \Delta n) f \neq 0. \tag{5.18}$$

This relation implies that the first and second operators of Eq. (5.16) cannot be interchanged (i.e., they are not *commutable*). However, we symmetrize the operators in Eq. (5.16) as

$$\phi(x, y, z + \Delta z) = \exp\left(-j\frac{\Delta z}{4\beta}\nabla_\perp^2\right) \exp(-j\chi) \exp\left(-j\frac{\Delta z}{4\beta}\nabla_\perp^2\right) \phi(x, y, z). \tag{5.19}$$

Although the reasons are not discussed here because of space limitations, Eq. (5.19) results in errors of the order of $(\Delta z)^3$. This is one of the reasons that the propagation step Δz in the FFT-BPM has to be small. It should also be pointed out that Eq. (5.16), which generates the unsymmetrical operators, results in larger errors: errors of the order of $(\Delta z)^2$ [2].

Now, let us discuss the physical meaning of each term in Eq. (5.19). When the refractive index is assumed to be uniform in the analysis region, χ is equal to zero. Thus, the wave equation Eq. (5.19) is reduced to

$$\phi(x, y, z + \Delta z) = \exp\left(-j\frac{\Delta z}{4\beta}\nabla_\perp^2\right) \cdot 1 \cdot \exp\left(-j\frac{\Delta z}{4\beta}\nabla_\perp^2\right) \phi(x, y, z)$$

$$= \exp\left(-j\frac{\Delta z}{2\beta}\nabla_\perp^2\right) \phi(x, y, z). \tag{5.20}$$

This implies that the operator

$$\exp\left(-j\frac{\Delta z}{2\beta}\nabla_\perp^2\right) \tag{5.21}$$

corresponds to the propagation over Δz in free space. Therefore, the first and the third terms of Eq. (5.19) correspond to the free-space propagation of the light over $\Delta z/2$. Thus, Eq. (5.19) implies that the wave function at $z + \Delta z$ can be obtained by first advancing the wave function at z by $\Delta z/2$ in free space, then giving a phase shift $(-\chi)$ due to a phase-shift lens, and finally advancing the wave function by another $\Delta z/2$ in free space.

Next, we will obtain the explicit expression of the free-space propagation operator (5.21) by actually applying it to the wave function.

The discrete Fourier transform (i.e., the spectral domain wave function) is

$$\tilde{\phi}_{mn}(z) = \sum_{i=0}^{M-1}\sum_{h=0}^{N-1} \phi(x,y,z)\exp\left[-j2\pi\left(\frac{mx}{X}+\frac{ny}{Y}\right)\right], \tag{5.22}$$

where

$$x = \Delta x \cdot i, \quad y = \Delta y \cdot h, \quad X = \Delta x \cdot M, \quad Y = \Delta y \cdot N,$$
$$0 \leq i \leq M-1, \quad 0 \leq h \leq N-1, \tag{5.23}$$
$$-\frac{M}{2} \leq m \leq \frac{M}{2}-1, \quad -\frac{N}{2} \leq n \leq \frac{N}{2}-1.$$

Here, X and Y are the widths in the x and y directions. The inverse discrete Fourier transform, on the other hand, is

$$\phi(x,y,z) = \sum_{m=-M/2}^{M/2-1}\sum_{n=-N/2}^{N/2-1} \tilde{\phi}_{mn}(z)\exp\left[j2\pi\left(\frac{mx}{X}+\frac{ny}{Y}\right)\right], \tag{5.24}$$

where the coefficient $1/MN$ is omitted for simplicity.

According to the operator (5.21) and the above discussion, when the wave propagates over $\Delta z/2$ in free space, we get the following wave function at $z + \Delta z/2$:

$$\phi\left(x,y,z+\frac{\Delta z}{2}\right) = \exp\left(-j\frac{\Delta z}{4\beta}\nabla_\perp^2\right)\phi(x,y,z). \tag{5.25}$$

We can get the wave function at $z + \Delta z/2$ by replacing z by $z + \Delta z/2$ in Eq. (5.24):

$$\phi\left(x, y, z + \frac{\Delta z}{2}\right) = \sum_{m=-M/2}^{M/2-1} \sum_{n=-N/2}^{N/2-1} \tilde{\phi}_{mn}\left(z + \frac{\Delta z}{2}\right) \exp\left[j2\pi\left(\frac{mx}{X} + \frac{ny}{Y}\right)\right]. \tag{5.26}$$

On the other hand, substituting the right-hand side of Eq. (5.24) for $\phi(x, y, z)$ on the left-hand side of Eq. (5.25), we get

$$\phi\left(x, y, z + \frac{\Delta z}{2}\right) = \exp\left(-j\frac{\Delta z}{4\beta}\nabla_\perp^2\right) \sum_{m=-M/2}^{M/2-1} \sum_{n=-N/2}^{N/2-1} \tilde{\phi}_{mn}(z)$$
$$\times \exp\left[j2\pi\left(\frac{mx}{X} + \frac{ny}{Y}\right)\right]. \tag{5.27}$$

As the right-hand side of this equation can be rewritten as

$$\sum_{m=-M/2}^{M/2-1} \sum_{n=-N/2}^{N/2-1} \tilde{\phi}_{mn}(z) \exp\left(-j\frac{\Delta z}{4\beta}\nabla_\perp^2\right) \exp\left[j2\pi\left(\frac{mx}{X} + \frac{ny}{Y}\right)\right]$$
$$= \sum_{m=-M/2}^{M/2-1} \sum_{n=-N/2}^{N/2-1} \tilde{\phi}_{mn}(z) \exp\left\{j\frac{(2\pi)^2}{4\beta}\left[\left(\frac{m}{X}\right)^2 + \left(\frac{n}{Y}\right)^2\right]\Delta z\right\}$$
$$\times \exp\left[j2\pi\left(\frac{mx}{X} + \frac{ny}{Y}\right)\right],$$

we get another expression for the wave function at $z + \Delta z/2$:

$$\phi\left(x, y, z + \frac{\Delta z}{2}\right) = \sum_{m=-M/2}^{M/2-1} \sum_{n=-N/2}^{N/2-1} \tilde{\phi}_{mn}(z)$$
$$\times \exp\left\{j\frac{(2\pi)^2}{4\beta}\left[\left(\frac{m}{X}\right)^2 + \left(\frac{n}{Y}\right)^2\right]\Delta z\right\}$$
$$\times \exp\left[j2\pi\left(\frac{mx}{X} + \frac{ny}{Y}\right)\right]. \tag{5.28}$$

Since the wave functions $\phi(x, y, z + \Delta z/2)$ given in Eqs. (5.26) and (5.28) have to be equal to each other, we get the important relation

$$\tilde{\phi}_{mn}\left(z + \frac{\Delta z}{2}\right) = \tilde{\phi}_{mn}(z) \exp\left\{j\frac{(2\pi)^2}{4\beta}\left[\left(\frac{m}{X}\right)^2 + \left(\frac{n}{Y}\right)^2\right] \Delta z\right\}. \quad (5.29)$$

Equation (5.29) shows the relation between the spectral domain wave function $\tilde{\phi}_{mn}(z + \Delta z/2)$ at $z + \Delta z/2$ and the spectral domain wave function $\tilde{\phi}_{mn}(z)$ at z. The exponential term on the right-hand side of Eq. (5.29),

$$\exp\left\{j\frac{(2\pi)^2}{4\beta}\left[\left(\frac{m}{X}\right)^2 + \left(\frac{n}{Y}\right)^2\right] \Delta z\right\},$$

corresponds to the propagation over $\Delta z/2$ in free space. We also find that Eq. (5.28) is the inverse discrete Fourier transform of the function

$$\tilde{\phi}_{mn}(z) \exp\left\{j\frac{(2\pi)^2}{4\beta}\left[\left(\frac{m}{X}\right)^2 + \left(\frac{n}{Y}\right)^2\right] \Delta z\right\}.$$

From these discussions, we draw the conclusion that the application of the operator

$$\exp\left(j\frac{\Delta z}{4\beta}\nabla_\perp^2\right), \quad (5.30)$$

which corresponds to the propagation over $\Delta z/2$ in free space, to the space-domain wave function $\phi(x, y, z)$ at z is equivalent to application of the mathematical procedure

$$\mathfrak{T}^{-1} \exp\left(j\frac{1}{4\beta}(k_x^2 + k_y^2) \Delta z\right) \mathfrak{T} \quad (5.31)$$

to the space-domain wave function $\phi(x, y, z)$. Here, the symbols $\mathfrak{T}$ and $\mathfrak{T}^{-1}$ respectively represent the discrete Fourier transform and the inverse discrete Fourier transform. The variables k_x and k_y are expressed as

$$k_x = \frac{2\pi m}{X} \quad \text{and} \quad k_y = \frac{2\pi n}{Y}. \quad (5.32)$$

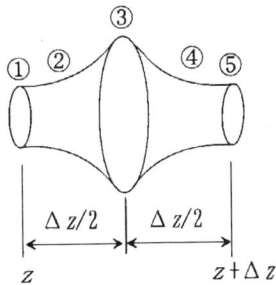

FIGURE 5.1. Calculation of one period in the FFT-BPM.

Thus, the FFT-BPM calculation procedures for a period Δz can be summarized as follows, where steps 1–5 correspond to the labels in Fig. 5.1:

1. At the propagation position z, calculate the spectral domain wave function $\tilde{\phi}_{mn}(z)$ in the Fourier transform domain by taking the Fourier transform of the space-domain wave function $\phi(x, y, z)$.
2. To get the transformed wave function $\tilde{\phi}_{mn}(z + \Delta z/2)$ at $z + \Delta z/2$, multiply

$$\exp\left\{j\frac{(2\pi)^2}{4\beta}\left[\left(\frac{mx}{X}\right)^2 + \left(\frac{ny}{Y}\right)^2\right]\Delta z\right\} \quad (5.33)$$

by the spectral domain wave function $\tilde{\phi}_{mn}(z)$ obtained in step 1. This multiplication corresponds to the propagation over the distance $\Delta z/2$ in free space.
3. Taking the inverse Fourier transform of the spectral domain wave function $\tilde{\phi}_{mn}(z + \Delta z/2)$ obtained in step 2, obtain the space-domain wave function $\phi(x, y, z + \Delta z/2)$ just in front of the phase-shift lens. Then, multiplying the phase-shift term $\exp(-j\chi)$ due to the phase-shift lens by the space-domain wave function $\phi(x, y, z + \Delta z/2)$, obtain the space-domain wave function just after the phase-shift lens:

$$\exp(-j\chi)\phi\left(x, y, z + \frac{\Delta z}{2}\right). \quad (5.34)$$

4. Taking the Fourier transform of the space-domain wave function just after the phase-shift lens and multiplying it by

$$\exp\left\{ j\frac{(2\pi)^2}{4\beta}\left[\left(\frac{mx}{X}\right)^2+\left(\frac{ny}{Y}\right)^2\right]\Delta z\right\}, \qquad (5.35)$$

corresponding to the propagation over $\Delta z/2$ in free space, obtain the spectral domain wave function $\tilde{\phi}_{mn}(z+\Delta z)$ at $z+\Delta z$.

5. When the space-domain wave function $\phi(x,y,z+\Delta z)$ at $z+\Delta z$ is necessary, take the inverse Fourier transform of the spectral domain wave function $\tilde{\phi}_{mn}(z+\Delta z)$ obtained in step 4.

Repeating steps 1–5, we can get the space-domain wave function at the target propagation position. It should be noted that if the space-domain wave function at each $z+\Delta z$ is not necessary, one should return directly to step 2 from step 4 and repeat steps 2–4.

5.1.3 Wide-Angle Formulation

Up to this stage (i.e., in the Fresnel approximation), the second derivative of the slowly varying envelope function ϕ with respect to z has been neglected. Now, we return to the wide-angle wave equation (5.9), which contains the second derivative. Readers who feel they have no need for such a discussion of the wide-angle formulation for the FFT-BPM can skip this section.

Substituting the wave function given by Eq. (5.12) into Eq. (5.9) and dividing both sides by ϕ, we get

$$2j\beta\gamma - \gamma^2 = \nabla_\perp^2 + k_0^2(n^2 - n_{\text{eff}}^2).$$

Therefore

$$\gamma^2 - 2j\beta\gamma + \nabla_\perp^2 + k_0^2(n^2 - n_{\text{eff}}^2) = 0$$

and

$$\gamma = j\beta \pm \sqrt{-\beta^2 - \{\nabla_\perp^2 + k_0^2(n^2 - n_{\text{eff}}^2)\}}$$
$$= j\beta \pm j\sqrt{\beta^2 + \{\nabla_\perp^2 + k_0^2(n^2 - n_{\text{eff}}^2)\}}. \qquad (5.36)$$

5.1 FAST FOURIER TRANSFORM BEAM PROPAGATION METHOD

Since the wave is supposed to propagate in the $+z$ direction, the minus sign in Eq. (5.36) should be used so that

$$\gamma = j\left(\beta - \sqrt{\beta^2 + \{\nabla_\perp^2 + k_0^2(n^2 - n_{\text{eff}}^2)\}}\right). \tag{5.37}$$

Substituting this γ into Eq. (5.12), we get the wave function at z:

$$\phi(x, y, z) = A(x, y) \exp\left[j\left(\beta - \sqrt{\beta^2 + \{\nabla_\perp^2 + k_0^2(n^2 - n_{\text{eff}}^2)\}}\right)z\right]. \tag{5.38}$$

Finally, we get the wave function at $z + \Delta z$:

$$\phi(x, y, z + \Delta z) = \exp\left[j\left(\beta - \sqrt{\beta^2 + \{\nabla_\perp^2 + k_0^2(n^2 - n_{\text{eff}}^2)\}}\right)\Delta z\right]\phi(x, y, z). \tag{5.39}$$

Now, we expand the second term inside the exponential function of Eq. (5.39):

$$(\beta^2 + \{\nabla_\perp^2 + k_0^2(n^2 - n_{\text{eff}}^2)\})^{1/2}$$

$$= (\beta^2 + \nabla_\perp^2)^{1/2}\left\{1 + \frac{k_0^2(n^2 - n_{\text{eff}}^2)}{\beta^2 + \nabla_\perp^2}\right\}^{1/2}$$

$$= (\beta^2 + \nabla_\perp^2)^{1/2}\left\{1 + \frac{1}{2}\frac{k_0^2(n^2 - n_{\text{eff}}^2)}{\beta^2 + \nabla_\perp^2}\right\}$$

$$(\because k_0^2(n^2 - n_{\text{eff}}^2) \ll \beta^2 + \nabla_\perp^2)$$

$$= (\beta^2 + \nabla_\perp^2)^{1/2} + \tfrac{1}{2}k_0^2(n^2 - n_{\text{eff}}^2)(\beta^2 + \nabla_\perp^2)^{-1/2}. \tag{5.40}$$

The Laplacian $\nabla_\perp^2$ in the lateral directions corresponds to the square of the wave numbers in the lateral directions. The wave numbers in the lateral directions are much smaller than the wave number in the propagation direction (i.e., the propagation constant). Thus, the assumption $\nabla_\perp^2 \ll \beta^2$ is concluded to be valid. In addition, taking $(n^2 - n_{\text{eff}}^2) = (n + n_{\text{eff}})(n - n_{\text{eff}}) \approx 2n_{\text{eff}}(n - n_{\text{eff}})$ and $\beta = k_0 n_{\text{eff}}$ into consideration, we can simplify the right-hand side of Eq. (5.40) to

$$(\beta^2 + \nabla_\perp^2)^{1/2} + k_0^2 \frac{n_{\text{eff}}}{\beta}(n - n_{\text{eff}}) = (\beta^2 + \nabla_\perp^2)^{1/2} + k_0(n - n_{\text{eff}}). \tag{5.41}$$

Substituting Eq. (5.41) into Eq. (5.39), we get

$$\phi(x, y, z + \Delta z)$$
$$= \exp\left[j\left(\beta - \sqrt{\beta^2 + \{\nabla_\perp^2 + k_0^2(n^2 - n_{\text{eff}}^2)\}}\right)\Delta x\right]\phi(x, y, z)$$
$$= \exp(j\{\beta - (\beta^2 + \nabla_\perp^2)^{1/2} - k_0(n - n_{\text{eff}})\}\Delta z)\phi(x, y, z)$$
$$= \exp[-jk_0(n - n_{\text{eff}})\Delta z]\exp(-j\{(\beta^2 + \nabla_\perp^2)^{1/2} - \beta\}\Delta z)\phi(x, y, z). \quad (5.42)$$

Since the propagation constant β has the order of the inverse of the wavelength, it is a large number. Inside the exponential function, $(\beta^2 + \nabla_\perp^2)^{1/2}$, which is also the large number, is subtracted by β. This numerical process can therefore cause large round-off errors. To reduce the errors, we modify the term inside the exponential function:

$$(\beta^2 + \nabla_\perp^2)^{1/2} - \beta = \frac{\{(\beta^2 + \nabla_\perp^2)^{1/2} - \beta\}\{(\beta^2 + \nabla_\perp^2)^{1/2} + \beta\}}{(\beta^2 + \nabla_\perp^2)^{1/2} + \beta}$$
$$= \frac{\nabla_\perp^2}{(\beta^2 + \nabla_\perp^2)^{1/2} + \beta}.$$

Symmetrizing the operators as we do in the wave equation (5.19) based on the Fresnel approximation, we finally get

$$\phi(x, y, z + \Delta z)$$
$$= \exp[-jk_0(n - n_{\text{eff}})\Delta z]\exp\left(-j\Delta z \frac{\nabla_\perp^2}{(\beta^2 + \nabla_\perp^2)^{1/2} + \beta}\right)\phi(x, y, z)$$
$$= \exp\left(-j\frac{\Delta z}{2}\frac{\nabla_\perp^2}{(\beta^2 + \nabla_\perp^2)^{1/2} + \beta}\right)\exp(-j\chi)$$
$$\times \exp\left(-j\frac{\Delta z}{2}\frac{\nabla_\perp^2}{(\beta^2 + \nabla_\perp^2)^{1/2} + \beta}\right)\phi(x, y, z). \quad (5.43)$$

This is a wide-angle formulation, so the operator for the propagation over $\Delta z/2$ in the Fresnel approximation corresponds to that in the wide-angle formulation as follows:

$$\underbrace{\exp\left(-j\frac{\Delta z}{4\beta}\nabla_\perp^2\right)}_{\text{Fresnel approximation}} \leftrightarrow \underbrace{\exp\left(-j\frac{\Delta z}{2}\frac{\nabla_\perp^2}{(\beta^2 + \nabla_\perp^2)^{1/2} + \beta}\right)}_{\text{Wide-angle formulation}}. \quad (5.44)$$

The calculation procedures for the beam propagation based on the wide-angle formulation are exactly the same as those for the beam propagation method based on the Fresnel approximation.

5.1.4 Analytical Boundaries

To reduce the reflections at analysis windows, we need some artificial boundary conditions. Since the simple transparent boundary condition (TBC) that is normally used in the FD-BPM (and which will be discussed in a later section) cannot be used in the FFT-BPM, some other artificial boundary conditions using complex refractive index materials or window functions have to be used to make the propagating fields decay properly near the edges of the analysis window. These artificial boundaries in the FFT-BPM usually require some experiences in optimizing the parameters to minimize the reflections.

5.1.5 Further Investigation

The free-space propagation operators for the Fresnel approximation and for the wide-angle formulation that were given in expression (5.44) are further examined in this section.

According to Eqs. (5.12) and (5.24), the slowly varying envelope function $\phi(x, z)$ in a 2D case is expressed as

$$\phi(x, z) = \sum_m \phi_m(x, z), \tag{5.45}$$

where

$$\phi_m(x, z) = \tilde{\phi}_m \exp(jk_x x) \exp(\gamma z) \tag{5.46}$$

and

$$k_x = \frac{2\pi m}{X}. \tag{5.47}$$

Here, X is the total width of the analysis region and m is an integer.

First, let us verify that the Fresnel approximation is equivalent to the para-axial approximation by using a 1D case, where we reduce the

178 BEAM PROPAGATION METHODS

derivative operator $\nabla_\perp^2$ in Eq. (5.44) to the square of a wave number in a lateral direction, k_x^2. The free-space propagation operators for the Fresnel approximation and the wide-angle formulation are respectively

$$\exp\left(j\frac{\Delta z}{4\beta}k_x^2\right), \tag{5.48}$$

$$\exp\left(j\frac{\Delta z}{2}\frac{k_x^2}{(\beta^2 - k_x^2)^{1/2} + \beta}\right). \tag{5.49}$$

Thus, the phase term φ_F for the Fresnel approximation and the phase term φ_W for the wide-angle formulation are

$$\varphi_F = \frac{\Delta z}{4\beta}k_x^2, \tag{5.50}$$

$$\varphi_W = \frac{\Delta z}{2}\frac{k_x^2}{(\beta^2 - k_x^2)^{1/2} + \beta}. \tag{5.51}$$

Equation (5.4) can be used to express β as a function of the reference index n_{eff}. Assuming that the reference index n_{eff} is equal to the effective index, we can, as shown in Fig. 5.2, reduce β to the z-directed component of the wave number k of a whole wave (i.e., the propagation constant).

Since k and β are generally much larger than the x-directed wave number k_x, the approximation $k \approx \beta$ is valid. Thus, k_x can be approximated as

$$k_x = k\sin\theta \approx \beta\sin\theta. \tag{5.52}$$

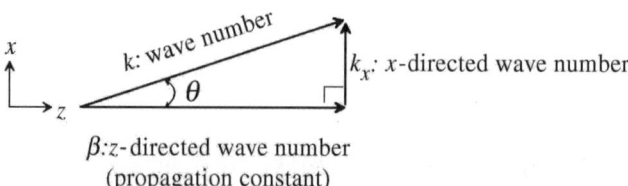

FIGURE 5.2. Relation between k, k_x, and β. Here, θ is the propagation angle.

5.1 FAST FOURIER TRANSFORM BEAM PROPAGATION METHOD

Substituting Eq. (5.52) into Eqs. (5.50) and Eq. (5.51), we can reduce the phase term φ_F for the Fresnel approximation and the phase term φ_W for the wide-angle formulation to

$$\varphi_F = \frac{\Delta z}{4\beta}\beta^2 \sin^2\theta = \frac{\Delta z}{4\beta}4\beta^2 \sin^2\frac{\theta}{2}\cos^2\frac{\theta}{2}, \tag{5.53}$$

$$\varphi_W = \frac{\Delta z}{2}\frac{k_x^2}{(\beta^2 - k_x^2)^{1/2} + \beta} = \frac{\Delta z}{2}\frac{\beta^2 \sin^2\theta}{(\beta^2 - \beta^2\sin^2\theta)^{1/2} + \beta}$$

$$= \frac{\Delta z}{2}\frac{\beta^2 4\sin^2(\theta/2)\cos^2(\theta/2)}{\beta(\cos\theta + 1)} = \frac{\Delta z}{2}\frac{\beta^2 4\sin^2(\theta/2)\cos^2(\theta/2)}{2\beta\cos^2(\theta/2)}$$

$$= \frac{\Delta z}{4\beta}4\beta^2 \sin^2\frac{\theta}{2}. \tag{5.54}$$

Comparing Eq. (5.53) with Eq. (5.54), we find that the Fresnel approximation is a good approximation for the wide-angle formulation only when the propagation angles are so small that $\cos^2(\theta/2)$ can be considered to be nearly equal to 1.

Next, we discuss the discretization width in the lateral directions. Since the free-space propagation operator (5.48) for the Fresnel approximation is purely imaginary, it satisfies the unitary condition. The free-space propagation operator (5.49) for the wide-angle formulation, on the other hand, has to satisfy

$$\beta^2 \geq k_x^2 \tag{5.55}$$

if the unitary condition is to be satisfied. That is, when

$$(k_0 n_{\text{eff}})^2 \geq \left(\frac{2\pi m}{X}\right)^2 \tag{5.56}$$

is not satisfied, the denominator of the argument of the operator (5.49) becomes complex. Thus, the free-space propagation operator (5.49) itself also becomes complex and does not satisfy the unitary condition. This causes the power of the propagating optical wave to dissipate. Assuming the maximum number of m to be $M/2$, we can rewrite condition (5.56) as

$$(k_0 n_{\text{eff}})^2 \geq \left(\frac{2\pi}{X}\frac{M}{2}\right)^2 \geq \left(\frac{\pi}{X/M}\right)^2 \geq \left(\frac{\pi}{\Delta x}\right)^2, \tag{5.57}$$

where $\Delta x = X/M$ is the discretization width. Since the relation $k_0 = 2\pi/\lambda_0$ holds for the wave number in a vacuum (λ_0 is the wavelength in the vacuum), Eq. (5.57) implies that the discretization width Δx in the lateral direction has to satisfy the condition

$$\frac{\lambda_0}{2n_{\text{eff}}} \leq \Delta x \qquad (5.58)$$

if the power of the beam is to be conserved. The condition (5.58) implies that a very small discretization cannot be used in the wide-angle FFT-BPM. For 3D cases, the conditions (5.57) and (5.58) are modified to

$$(k_0 n_{\text{eff}})^2 \geq \left(\frac{2\pi}{X}\frac{M}{2}\right)^2 \times 2 \geq \left(\frac{\pi}{X/M}\right)^2 \times 2 \geq \left(\frac{\pi}{\Delta x}\right) \times 2 \qquad (5.59)$$

and

$$\frac{\lambda_0}{\sqrt{2}n_{\text{eff}}} \leq \Delta x, \qquad (5.60)$$

where we assume that $\Delta x = \Delta y$.

5.2 FINITE-DIFFERENCE BEAM PROPAGATION METHOD

The FD-BPM is very powerful and has been widely used for optical waveguide design. Of the various FD-BPMs that have been developed, the one with the implicit scheme developed by Chung and Dagli [4] is state-of-the-art from the viewpoints of accuracy, numerical efficiency, and stability. Its unconditional stability is particularly advantageous not only because it allows us to use the method in actual design without danger of diversion but also because it allows us to set the propagation step relatively large. In addition, the TBC [13], which is simple and requires no special experience to use, has been developed for the FD-BPM by Hadley [5]. A wide-angle scheme using Padé approximant operators [5, 6] has also been developed by him. These contributions greatly advanced the FD-BPM and have enabled it to be used even in the design of optical waveguides made of high-contrast-index materials, such as semiconductor optical waveguides.

5.2.1 Wave Equation

To clarify the formulation of the FD-BPM, we must first derive the wave equation. In this section, as in Chapter 2, we assume that the structure of the optical waveguide is uniform in the y direction. The necessary wave equations can be derived from the semivectorial wave equations given in Chapter 4, but let us further our understanding by deriving them from Maxwell's equations.

The component representations of Maxwell's equations

$$\nabla \times \mathbf{E} = -j\omega\mu_0 \mathbf{H}, \tag{5.61}$$

$$\nabla \times \mathbf{H} = j\omega\varepsilon_0\varepsilon_r \mathbf{E} \tag{5.62}$$

were given in Eqs. (2.5)–(2.10), where the relative permeability μ_r was assumed to be equal to 1. Since the structure in the y direction is assumed to be uniform, the derivatives with respect to y can be set to zero. Thus, Eqs. (2.5)–(2.10) are reduced to

$$-\frac{\partial E_y}{\partial z} = -j\omega\mu_0 H_x, \tag{5.63}$$

$$\frac{\partial E_x}{\partial z} - \frac{\partial E_z}{\partial x} = -j\omega\mu_0 H_y, \tag{5.64}$$

$$\frac{\partial E_y}{\partial x} = -j\omega\mu_0 H_z, \tag{5.65}$$

$$-\frac{\partial H_y}{\partial z} = j\omega\varepsilon_0\varepsilon_r E_x, \tag{5.66}$$

$$\frac{\partial H_x}{\partial z} - \frac{\partial H_z}{\partial x} = j\omega\varepsilon_0\varepsilon_r E_y, \tag{5.67}$$

$$\frac{\partial H_y}{\partial x} = j\omega\varepsilon_0\varepsilon_r E_z. \tag{5.68}$$

A. TE Mode Figure 5.3 shows the principal field components for the TE mode. Since both E_y and H_x are the principal fields, we need to have the wave equations for both. In the TE mode, as discussed in Section 2.1.1, the x- and z-directed electric field components and the y-directed magnetic field component are zero:

$$E_x = E_z = H_y = 0. \tag{5.69}$$

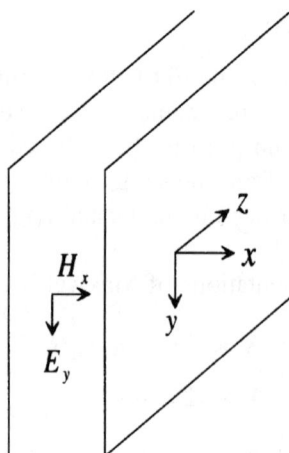

FIGURE 5.3. Principal field components for TE mode are E_y and H_x.

Substituting Eq. (5.69) into Eqs. (5.63)–(5.68), we obtain for the TE mode the equations

$$-\frac{\partial E_y}{\partial z} = -j\omega\mu_0 H_x, \tag{5.70}$$

$$\frac{\partial E_y}{\partial x} = -j\omega\mu_0 H_z, \tag{5.71}$$

$$\frac{\partial H_x}{\partial z} - \frac{\partial H_z}{\partial x} = j\omega\varepsilon_0\varepsilon_r E_y. \tag{5.72}$$

1. E_y Representation First, we derive the wave equation for the y-directed electric field component E_y. Substituting the x- and z-directed magnetic field components

$$H_x = \frac{1}{j\omega\mu_0}\frac{\partial E_y}{\partial z}, \tag{5.73}$$

$$H_z = -\frac{1}{j\omega\mu_0}\frac{\partial E_y}{\partial x} \tag{5.74}$$

obtained from Eqs. (5.70) and (5.71) into Eq. (5.72) yields the following wave equation for the principal electric field component E_y:

$$\frac{\partial^2 E_y}{\partial z^2} + \frac{\partial^2 E_y}{\partial x^2} + k_0^2 \varepsilon_r E_y = 0, \quad (5.75)$$

where $k_0^2 = \omega^2 \varepsilon_0 \mu_0$.

2. H_x Representation Next, we derive the wave equation for the x-directed magnetic field component H_x. Differentiating Eq. (5.72) with respect to z, we get

$$\frac{\partial^2 H_x}{\partial z^2} - \frac{\partial^2 H_z}{\partial z \, \partial x} = j\omega \varepsilon_0 \varepsilon_r \frac{\partial E_y}{\partial z}. \quad (5.76)$$

In deriving Eq. (5.76), we have assumed the variation of the relative permittivity ε_r along the propagation axis to be negligibly small. That is,

$$\frac{\partial \varepsilon_r}{\partial z} \approx 0. \quad (5.77)$$

This results in the approximation

$$\frac{\partial}{\partial z}(\varepsilon_r E_y) = \frac{\partial \varepsilon_r}{\partial z} E_y + \varepsilon_r \frac{\partial E_y}{\partial z} \approx \varepsilon_r \frac{\partial E_y}{\partial z}. \quad (5.78)$$

Thus, it should be noted that the approximation (5.77) is implicitly used in the BPM.

From a calculation of ∂Eq. (5.70)$/\partial x + \partial$Eq. (5.70)$/\partial z$ or the magnetic divergence equation

$$\nabla \cdot \mathbf{H} = 0, \quad (5.79)$$

we get

$$\frac{\partial H_x}{\partial x} + \frac{\partial H_z}{\partial z} = 0.$$

That is,

$$\frac{\partial H_z}{\partial z} = -\frac{\partial H_x}{\partial x}.$$

Substituting this equation and Eq. (5.70) into Eq. (5.76), we can eliminate H_z in Eq. (5.76). This results in the wave equation for the principal magnetic field component H_x:

$$\frac{\partial^2 H_x}{\partial z^2} + \frac{\partial^2 H_x}{\partial x^2} + k_0^2 \varepsilon_r H_x = 0. \tag{5.80}$$

B. TM Mode Figure 5.4 shows the principal field components for the TM mode. Since both E_x and H_y are principal fields, we need to have the wave equations for both. In the TE mode, as discussed in Section 2.1.2, the x- and z-directed magnetic field components and y-directed electric field component are zero:

$$H_x = H_z = E_y = 0. \tag{5.81}$$

Substituting Eq. (5.81) into Eqs. (5.63)–(5.68), we obtain for the TM mode the equations

$$\frac{\partial E_x}{\partial z} - \frac{\partial E_z}{\partial x} = -j\omega\mu_0 H_y, \tag{5.82}$$

$$-\frac{\partial H_y}{\partial z} = j\omega\varepsilon_0\varepsilon_r E_x, \tag{5.83}$$

$$\frac{\partial H_y}{\partial x} = j\omega\varepsilon_0\varepsilon_r E_z. \tag{5.84}$$

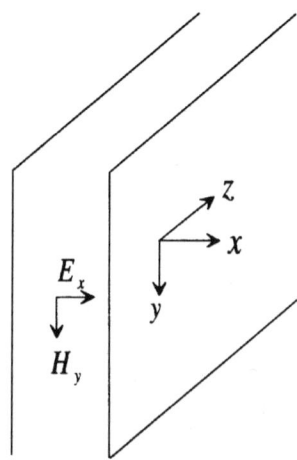

FIGURE 5.4. Principal field components for TM mode are E_x and H_y.

1. E_x Representation First, we derive the wave equation for the x-directed electric field component E_x. Differentiating Eq. (5.82) with respect to z, we get

$$\frac{\partial^2 E_x}{\partial z^2} - \frac{\partial^2 E_z}{\partial z\, \partial x} = -j\omega\mu_0 \frac{\partial H_y}{\partial z}. \tag{5.85}$$

Substituting Eq. (5.83) into Eq. (5.85), we can eliminate H_y and obtain

$$\frac{\partial^2 E_x}{\partial z^2} - \frac{\partial^2 E_z}{\partial z\, \partial x} = -j\omega\mu_0(-j\omega\varepsilon_0\varepsilon_r E_x) = -k_0^2 \varepsilon_r E_x. \tag{5.86}$$

From a calculation of ∂Eq. (5.83)$/\partial x + \partial$Eq. (5.84)$/\partial z$ or the divergence equation for the electric flux density

$$\nabla \cdot (\varepsilon_r \mathbf{E}) = 0, \tag{5.87}$$

we get

$$\frac{\partial}{\partial x}(\varepsilon_r E_x) + \frac{\partial}{\partial z}(\varepsilon_r E_z) = 0.$$

That is,

$$\frac{\partial E_z}{\partial z} = -\frac{1}{\varepsilon_r}\frac{\partial}{\partial x}(\varepsilon_r E_x).$$

Substituting this equation into Eq. (5.86), we get the wave equation for the principal electric field component E_x:

$$\frac{\partial^2 E_x}{\partial z^2} + \frac{\partial}{\partial x}\left(\frac{1}{\varepsilon_r}\frac{\partial}{\partial x}(\varepsilon_r E_x)\right) + k_0^2 \varepsilon_r E_x = 0. \tag{5.88}$$

2. H_y Representation Next, we derive the wave equation for the y-directed magnetic field component H_y. Substituting the x- and z-directed electric field components

$$E_x = -\frac{1}{j\omega\varepsilon_0\varepsilon_r}\frac{\partial H_y}{\partial z}, \tag{5.89}$$

$$E_z = \frac{1}{j\omega\varepsilon_0\varepsilon_r}\frac{\partial H_y}{\partial x}, \tag{5.90}$$

which are obtained from Eqs. (5.83) and (5.84), into Eq. (5.82) yields

$$\frac{\partial}{\partial z}\left(-\frac{1}{j\omega\varepsilon_0\varepsilon_r}\frac{\partial H_y}{\partial z}\right) - \frac{\partial}{\partial x}\left(\frac{1}{j\omega\varepsilon_0\varepsilon_r}\frac{\partial H_y}{\partial x}\right) = -j\omega\mu_0 H_y \tag{5.91}$$

and therefore

$$\frac{\partial}{\partial z}\left(\frac{1}{\varepsilon_r}\frac{\partial H_y}{\partial z}\right) + \frac{\partial}{\partial x}\left(\frac{1}{\varepsilon_r}\frac{\partial H_y}{\partial x}\right) = -\omega^2 \varepsilon_0 \mu_0 H_y = -k_0^2 H_y.$$

Making use of the approximation

$$\frac{\partial}{\partial z}\left(\frac{1}{\varepsilon_r}\frac{\partial H_y}{\partial z}\right) = -\frac{1}{\varepsilon_r^2}\frac{\partial \varepsilon_r}{\partial z}\frac{\partial H_y}{\partial z} + \frac{1}{\varepsilon_r}\frac{\partial^2 H_y}{\partial z^2} \approx \frac{1}{\varepsilon_r}\frac{\partial^2 H_y}{\partial z^2},$$

which follows from Eq. (5.77), we get the wave equation for the principal magnetic field component H_y:

$$\frac{\partial^2 H_y}{\partial z^2} + \varepsilon_r \frac{\partial}{\partial x}\left(\frac{1}{\varepsilon_r}\frac{\partial H_y}{\partial x}\right) + k_0^2 \varepsilon_r H_y = 0. \qquad (5.92)$$

5.2.2 FD-BPM Formulation

Next, let us discuss the FD-BPM formulation based on the implicit scheme developed by Chung and Dagli [4]. Since the discussions here are limited to 2D problems, the amount of memory required is not large. Thus, the equidistant discretization is used to ensure the second-order accuracy. Further improvement of accuracy has been achieved by Yamauchi et al. [7].

A. TE Mode The wave equation for the y-directed electric field $E_y(x, y, z)$ is

$$\frac{\partial^2 E_y}{\partial z^2} + \frac{\partial^2 E_y}{\partial x^2} + k_0^2 \varepsilon_r E_y = 0. \qquad (5.93)$$

As in the FFT-BPM, using the slowly varying envelope approximation, we divide the principal field $E_y(x, y, z)$ propagating in the z direction into the slowly varying envelope function $\phi(x, y, z)$ and the very fast oscillatory phase term $\exp(-j\beta z)$ as follows:

$$E_y(x, y, z) = \phi(x, y, z) \exp(-j\beta z), \qquad (5.94)$$

5.2 FINITE-DIFFERENCE BEAM PROPAGATION METHOD

where

$$\beta = n_{\text{eff}} k_0. \tag{5.95}$$

Here, k_0 is the wave number in the vacuum and n_{eff} is the reference index, for which the effective index is usually used. Substituting

$$\frac{\partial^2 E_y}{\partial z^2} = \frac{\partial^2 \phi}{\partial z^2} \exp(-j\beta z) - 2j\beta \frac{\partial \phi}{\partial z} \exp(-j\beta z) - \beta^2 \phi \exp(-j\beta z), \tag{5.96}$$

which is obtained from Eq. (5.94), into Eq. (5.93) and dividing both sides of the resultant equation by the exponential term $\exp(-j\beta z)$, we get

$$\frac{\partial^2 \phi}{\partial z^2} - 2j\beta \frac{\partial \phi}{\partial z} + \frac{\partial^2 \phi}{\partial x^2} + (k_0^2 n^2 - \beta^2)\phi = 0 \tag{5.97}$$

or

$$2j\beta \frac{\partial \phi}{\partial z} - \frac{\partial^2 \phi}{\partial z^2} = \frac{\partial^2 \phi}{\partial x^2} + k_0^2(\varepsilon_r - n_{\text{eff}}^2)\phi, \tag{5.98}$$

where we used the relation $\varepsilon_r = n^2$. As discussed in the section covering the FFT-BPM, Eq. (5.98) is the wide-angle formulation. When we assume that

$$\frac{\partial^2 \phi}{\partial z^2} = 0, \tag{5.99}$$

Eq. (5.98) is reduced to the Fresnel wave equation

$$2j\beta \frac{\partial \phi}{\partial z} = \frac{\partial^2 \phi}{\partial x^2} + k_0^2(\varepsilon_r - n_{\text{eff}}^2)\phi. \tag{5.100}$$

First, we discuss the FD expression for the Fresnel approximation. That for the wide-angle formulation will be covered in Section 5.3.

When we use the discretization of the x and z coordinates

$$x = p \, \Delta x, \tag{5.101}$$

$$z = l \, \Delta z, \tag{5.102}$$

where p and q are integers, the following notations are used for the wave function $\phi(x, z)$ and the relative permittivity $\varepsilon_r(x, z)$:

$$\phi(x, z) \to \phi_p^l, \tag{5.103}$$

$$\varepsilon_r(x, z) \to \varepsilon_r^l(p). \tag{5.104}$$

The next step is the discretization of the Fresnel wave equation (5.100). First, we discretize it in the x direction. The discretization number l, which corresponds to the z coordinate, will be discussed later. The first and the second terms on the right-hand side of Eq. (5.100) are expressed as

$$\frac{\partial^2 \phi}{\partial x^2} = \frac{1}{\Delta x}\left(\underbrace{\frac{\phi_{p+1} - \phi_p}{\Delta x}}_{\text{Difference center is } p+\frac{1}{2}} - \underbrace{\frac{\phi_p - \phi_{p-1}}{\Delta x}}_{\text{Difference center is } p-\frac{1}{2}} \right)$$

$$= \underbrace{\frac{\phi_{p+1} - 2\phi_p + \phi_{p-1}}{(\Delta x)^2}}_{\text{Difference center is } p} \tag{5.105}$$

and

$$k_0^2(\varepsilon_r - n_{\text{eff}}^2)\phi = \underbrace{k_0^2[\varepsilon_r(p) - n_{\text{eff}}^2]\phi_p}_{\text{Difference center is } p}. \tag{5.106}$$

Substituting Eqs. (5.105) and (5.106) into Eq. (5.100), we get

$$2j\beta \frac{\partial \phi_p}{\partial z} = \frac{\phi_{p+1} - 2\phi_p + \phi_{p-1}}{(\Delta x)^2} + k_0^2[\varepsilon_r(p) - n_{\text{eff}}^2]\phi_p$$

$$= \alpha_w \phi_{p-1} + \alpha_x \phi_p + \alpha_e \phi_{p+1} + k_0^2[\varepsilon_r(p) - n_{\text{eff}}^2]\phi_p.$$

Thus, the discretization of the wave equation (5.100) is

$$2j\beta \frac{\partial \phi_p}{\partial z} = \alpha_w \phi_{p-1} + \{\alpha_x + k_0^2[\varepsilon_r(p) - n_{\text{eff}}^2]\}\phi_p + \alpha_e \phi_{p+1}, \tag{5.107}$$

where we used the definitions

$$\alpha_w = \frac{1}{(\Delta x)^2}, \tag{5.108}$$

$$\alpha_e = \frac{1}{(\Delta x)^2}, \tag{5.109}$$

$$\alpha_x = -\frac{2}{(\Delta x)^2}. \tag{5.110}$$

The next step is the discretization of Eq. (5.107) with respect to z. Discretizing the left-hand side of Eq. (5.107) with respect to z, we get

$$2j\beta \frac{\phi_p^{l+1} - \phi_p^l}{\Delta z}. \tag{5.111}$$

It should be noted that, as shown in expression (5.111), the difference center of the left-hand side of Eq. (5.107) is the point $l+\frac{1}{2}$ midway between l and $l+1$. The difference center of the right-hand side of Eq. (5.107) discretized with respect to z should be $l+\frac{1}{2}$. Thus, we modify Eq. (5.107) to

$$2j\beta \frac{\phi_p^{l+1} - \phi_p^l}{\Delta z} = \tfrac{1}{2}[\alpha_w^l \phi_{p-1}^l + \{\alpha_x^l + k_0^2[\varepsilon_r^l(p) - n_{\text{eff}}^2]\}\phi_p^l + \alpha_e^l \phi_{p+1}^l]$$
$$+ \tfrac{1}{2}[\alpha_w^{l+1} \phi_{p-1}^{l+1} + \{\alpha_x^{l+1} + k_0^2[\varepsilon_r^{l+1}(p) - n_{\text{eff}}^2]\}\phi_p^{l+1}$$
$$+ \alpha_e^{l+1} \phi_{p+1}^{l+1}]. \tag{5.112}$$

Rewriting this equation so that the terms on the left- and right-hand sides respectively contain $l+1$ and l, we get

$$-\frac{\alpha_w^{l+1}}{2}\phi_{p-1}^{l+1} + \left\{-\frac{\alpha_x^{l+1}}{2} + \frac{2j\beta}{\Delta z} - \frac{1}{2}k_0^2[\varepsilon_r^{l+1}(p) - n_{\text{eff}}^2]\right\}\phi_p^{l+1} - \frac{\alpha_e^{l+1}}{2}\phi_{p+1}^{l+1}$$
$$= \frac{\alpha_w^l}{2}\phi_{p-1}^l + \left\{\frac{\alpha_x^l}{2} + \frac{2j\beta}{\Delta z} + \frac{1}{2}k_0^2[\varepsilon_r^l(p) - n_{\text{eff}}^2]\right\}\phi_p^l + \frac{\alpha_e^l}{2}\phi_{p+1}^l.$$

Multiplying both sides of this equation by 2, we get the FD expression for the TE mode:

$$-\alpha_w^{l+1}\phi_{p-1}^{l+1} + \left\{-\alpha_x^{l+1} + \frac{4j\beta}{\Delta z} - k_0^2[\varepsilon_r^{l+1}(p) - n_{\text{eff}}^2]\right\}\phi_p^{l+1} - \alpha_e^{l+1}\phi_{p+1}^{l+1}$$

$$= \alpha_w^l\phi_{p-1}^l + \left\{\alpha_x^l + \frac{4j\beta}{\Delta z} + k_0^2[\varepsilon_r^l(p) - n_{\text{eff}}^2]\right\}\phi_p^l + \alpha_e^l\phi_{p+1}^l. \quad (5.113)$$

Although in Eq. (5.113) we use the wave number in a vacuum, k_0, for ease of understanding, it is recommended that in actual programming the coordinates (i.e., x, y, and z) be multiplied by k_0 and the propagation constant β be divided by k_0 in order to reduce the round-off errors. The resultant formulation corresponds to dividing both sides of Eq. (5.113) by k_0^2.

B. TM Mode The wave equation (5.92) for the principal magnetic field component H_y is

$$\frac{\partial^2 H_y}{\partial z^2} + \varepsilon_r \frac{\partial}{\partial x}\left(\frac{1}{\varepsilon_r}\frac{\partial H_y}{\partial x}\right) + k_0^2 \varepsilon_r H_y = 0. \quad (5.114)$$

As in Eq. (5.94), the principal field $H_y(x, y, z)$ propagating in the z direction is divided into the slowly varying envelope function $\phi(x, y, z)$ and the very fast oscillatory phase term $\exp(-j\beta z)$ as follows:

$$H_y(x, y, z) = \phi(x, y, z)\exp(-j\beta z). \quad (5.115)$$

Substituting the second derivative of Eq. (5.115), which corresponds to Eq. (5.96), into Eq. (5.114) and dividing the resultant equation by the exponential term $\exp(-j\beta z)$, we get for the TM mode the wide-angle wave equation

$$2j\beta\frac{\partial \phi}{\partial z} - \frac{\partial^2 \phi}{\partial z^2} = \varepsilon_r \frac{\partial}{\partial x}\left(\frac{1}{\varepsilon_r}\frac{\partial \phi}{\partial x}\right) + k_0^2(\varepsilon_r - n_{\text{eff}}^2)\phi. \quad (5.116)$$

When we assume that

$$\frac{\partial^2 \phi}{\partial z^2} = 0, \quad (5.117)$$

we get the Fresnel wave equation

$$2j\beta \frac{\partial \phi}{\partial z} = \varepsilon_r \frac{\partial}{\partial x}\left(\frac{1}{\varepsilon_r}\frac{\partial \phi}{\partial x}\right) + k_0^2(\varepsilon_r - n_{\text{eff}}^2)\phi. \tag{5.118}$$

To discretize the wave equation (5.118), we express the x and z coordinates, the wave function $\phi(x, z)$, and the relative permittivity $\varepsilon_r(x, z)$ as follows:

$$x = p\,\Delta x, \tag{5.119}$$

$$z = l\,\Delta z, \tag{5.120}$$

$$\phi(x, z) = \phi_p^l, \tag{5.121}$$

$$\varepsilon_r(x, z) = \varepsilon_r^l(p). \tag{5.122}$$

We first discretize the Fresnel wave equation (5.118) in the x direction. The first term on the right-hand side of Eq. (5.118) is discretized as follows:

$$\varepsilon_r \frac{\partial}{\partial x}\left(\frac{1}{\varepsilon_r}\frac{\partial \phi}{\partial x}\right)$$

$$= \varepsilon_r(p)\frac{1}{\Delta x}\left(\underbrace{\frac{1}{\varepsilon_r(p+1/2)}\frac{\phi_{p+1} - \phi_p}{\Delta x}}_{\text{Difference center is } p+1/2} - \underbrace{\frac{1}{\varepsilon_r(p-1/2)}\frac{\phi_p - \phi_{p_1}}{\Delta x}}_{\text{Difference center is } p-1/2}\right), \tag{5.123}$$

where

$$\varepsilon_r\left(p + \frac{1}{2}\right) \approx \frac{\varepsilon_r(p+1) + \varepsilon_r(p)}{2}, \tag{5.124}$$

$$\varepsilon_r\left(p - \frac{1}{2}\right) \approx \frac{\varepsilon_r(p) + \varepsilon_r(p-1)}{2}. \tag{5.125}$$

Thus, we get

$$\varepsilon_r \frac{\partial}{\partial x}\left(\frac{1}{\varepsilon_r}\frac{\partial \phi}{\partial x}\right)$$

$$= \varepsilon_r(p)\frac{1}{\Delta x}\left(\underbrace{\frac{2}{\varepsilon_r(p+1)+\varepsilon_r(p)}\frac{\phi_{p+1}-\phi_p}{\Delta x}}_{\text{Difference center is } p+1/2} - \underbrace{\frac{2}{\varepsilon_r(p)+\varepsilon_r(p-1)}\frac{\phi_p-\phi_{p-1}}{\Delta x}}_{\text{Difference center is } p-1/2}\right)$$

$$= \frac{2\varepsilon_r(p)}{\varepsilon_r(p+1)+\varepsilon_r(p)}\frac{\phi_{p+1}-\phi_p}{(\Delta x)^2} - \frac{2\varepsilon_r(p)}{\varepsilon_r(p)+\varepsilon_r(p-1)}\frac{\phi_p-\phi_{p-1}}{(\Delta x)^2}$$

$$= \frac{1}{(\Delta x)^2}\frac{2\varepsilon_r(p)}{\varepsilon_r(p+1)+\varepsilon_r(p)}\phi_{p+1} + \frac{1}{(\Delta x)^2}\frac{2\varepsilon_r(p)}{\varepsilon_r(p)+\varepsilon_r(p-1)}\phi_{p-1}$$

$$- \left(\frac{2\varepsilon_r(p)}{\varepsilon_r(p+1)+\varepsilon_r(p)} + \frac{2\varepsilon_r(p)}{\varepsilon_r(p)+\varepsilon_r(p-1)}\right)\frac{1}{(\Delta x)^2}\phi_p$$

$$= \alpha_w \phi_{p-1} + \alpha_x \phi_p + \alpha_e \phi_{p+1}, \tag{5.126}$$

where

$$\alpha_w = \frac{1}{(\Delta x)^2}\frac{2\varepsilon_r(p)}{\varepsilon_r(p)+\varepsilon_r(p-1)}, \tag{5.127}$$

$$\alpha_e = \frac{1}{(\Delta x)^2}\frac{2\varepsilon_r(p)}{\varepsilon_r(p)+\varepsilon_r(p+1)}, \tag{5.128}$$

$$\alpha_x = -\frac{1}{(\Delta x)^2}\frac{2\varepsilon_r(p)}{\varepsilon_r(p)+\varepsilon_r(p-1)} - \frac{1}{(\Delta x)^2}\frac{2\varepsilon_r(p)}{\varepsilon_r(p)+\varepsilon_r(p+1)}$$

$$= -\alpha_e - \alpha_w. \tag{5.129}$$

We also get

$$k_0^2(\varepsilon_r - n_{\text{eff}}^2)\phi = \underbrace{k_0^2(\varepsilon_r(p) - n_{\text{eff}}^2)\phi_p}_{\text{Difference center is } p}. \tag{5.130}$$

Substituting Eqs. (5.126) and (5.130) into Eq. (5.118), we get

$$2j\beta\frac{\partial \phi}{\partial z} = \alpha_w \phi_{p-1} + \alpha_x \phi_p + \alpha_e \phi_{p+1} + k_0^2[\varepsilon_r(p) - n_{\text{eff}}^2]\phi_p$$

and therefore

$$2j\beta\frac{\partial \phi}{\partial z} = \alpha_w \phi_{p-1} + \{\alpha_x + k_0^2[\varepsilon_r(p) - n_{\text{eff}}^2]\}\phi_p + \alpha_e \phi_{p+1}. \tag{5.131}$$

The next step is the discretization of Eq. (5.131) with respect to z. Discretizing the left-hand side of Eq. (5.131) with respect to z, we get

$$2j\beta \frac{\phi_p^{l+1} - \phi_p^l}{\Delta z}. \qquad (5.132)$$

As with expression (5.111), since the difference center of the left-hand side of Eq. (5.131) is the point $l + \frac{1}{2}$ midway between l and $l+1$ as shown in Eq. (5.131), the difference center of the right-hand side of Eq. (5.131) should be $l + \frac{1}{2}$. Using the same procedure, we used for the TE mode, we get the FD expression

$$-\alpha_w^{l+1}\phi_{p-1}^{l+1} + \left\{-\alpha_x^{l+1} + \frac{4j\beta}{\Delta z} - k_0^2[\varepsilon_r^{l+1}(p) - n_{\text{eff}}^2]\right\}\phi_p^{l+1} - \alpha_e^{l+1}\phi_{p+1}^{l+1}$$
$$= \alpha_w^l \phi_{p-1}^l + \left\{\alpha_x^l + \frac{4j\beta}{\Delta z} + k_0^2[\varepsilon_r^l(p) - n_{\text{eff}}^2]\phi_p^l + \alpha_e^l \phi_{p+1}^l\right\}. \qquad (5.133)$$

5.2.3 Nonequidistant Discretization Scheme

In the above discussions, we first discretized the Fresnel wave equation in the x direction as

$$2j\beta \frac{\partial \phi}{\partial z} = \alpha_w \phi_{p-1} + \{\alpha_x + k_0^2[\varepsilon_r(p) - n_{\text{eff}}^2]\}\phi_p + \alpha_e \phi_{p+1}. \qquad (5.134)$$

Then, discretizing Eq. (5.134) in the z direction, we got the FD expression for the Fresnel wave equation for a 1D nonequidistant discretization, which is shown in Fig. 5.5:

$$-\alpha_w^{l+1}\phi_{p-1}^{l+1} + \left\{-\alpha_x^{l+1} + \frac{4j\beta}{\Delta z} - k_0^2[\varepsilon_r^{l+1}(p) - n_{\text{eff}}^2]\right\}\phi_p^{l+1} - \alpha_e^{l+1}\phi_{p+1}^{l+1}$$
$$= \alpha_w^l \phi_{p-1}^l + \left\{\alpha_x^l + \frac{4j\beta}{\Delta z} + k_0^2[\varepsilon_r^l(p) - n_{\text{eff}}^2]\right\}\phi_p^l + \alpha_e^l \phi_{p+1}^l,$$
$$(5.135)$$

where the difference centers of both sides of Eq. (5.134) are the same in the z direction.

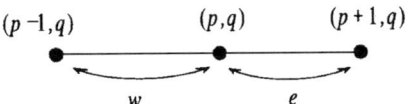

FIGURE 5.5. One-dimensional nonequidistant discretization.

a. TE mode—E_y representation and H_x representation:

$$\alpha_w = \frac{2}{w(e+w)}, \tag{5.136}$$

$$\alpha_e = \frac{2}{e(e+w)}, \tag{5.137}$$

$$\alpha_x = -\frac{2}{ew}$$

$$= -\frac{4}{ew} + \alpha_e + \alpha_w$$

$$= -\alpha_e - \alpha_w. \tag{5.138}$$

b. TM mode:

E_x representation:

$$\alpha_w = \frac{2}{w(e+w)} \frac{2\varepsilon_r(p-1)}{\varepsilon_r(p) + \varepsilon_r(p-1)}, \tag{5.139}$$

$$\alpha_e = \frac{2}{e(e+w)} \frac{2\varepsilon_r(p+1)}{\varepsilon_r(p) + \varepsilon_r(p+1)}, \tag{5.140}$$

$$\alpha_x = -\frac{2}{ew} - \frac{2}{w(e+w)} \frac{\varepsilon_r(p) - \varepsilon_r(p-1)}{\varepsilon_r(p) + \varepsilon_r(p-1)}$$

$$- \frac{2}{e(e+w)} \frac{\varepsilon_r(p) - \varepsilon_r(p+1)}{\varepsilon_r(p) + \varepsilon_r(p+1)}$$

$$= -\frac{4}{ew} + \alpha_e + \alpha_w. \tag{5.141}$$

H_y representation:

$$\alpha_w = \frac{2}{w(e+w)} \frac{2\varepsilon_r(p)}{\varepsilon_r(p) + \varepsilon_r(p-1)}, \tag{5.142}$$

$$\alpha_e = \frac{2}{e(e+w)} \frac{2\varepsilon_r(p)}{\varepsilon_r(p) + \varepsilon_r(p+1)}, \tag{5.143}$$

$$\alpha_x = -\frac{2}{w(e+w)} \frac{2\varepsilon_r(p)}{\varepsilon_r(p) + \varepsilon_r(p-1)}$$

$$- \frac{2}{e(e+w)} \frac{2\varepsilon_r(p)}{\varepsilon_r(p) + \varepsilon_r(p+1)}$$

$$= -\alpha_e + \alpha_w. \tag{5.144}$$

5.2.4 Stability Condition

Up to this stage, in the discretization of Eqs. (5.100) and (5.118) in the z direction, the difference centers were assumed to be the same (i.e., the point $z + \Delta z/2$ midway between z and $z + \Delta z$) for both the left- and the right-hand sides.

Here, we further discuss the influence of the difference center of the right-hand side in the z direction on the stability along the beam propagation by using the procedure discussed in Ref. [11].

The Fresnel equation is

$$2j\beta \frac{\partial \phi(x,z)}{\partial z} = \frac{\partial^2 \phi(x,z)}{\partial x^2} + k_0^2(\varepsilon_r - n_{\text{eff}}^2)\phi(x,z). \tag{5.145}$$

For simplicity, uniform media are assumed here. Since the reference index n_{eff} can be set to the refractive index $\sqrt{\varepsilon_r}$ of the medium, we can assume $n_{\text{eff}}^2 = \varepsilon_r$. Thus, the wave equation (5.145) can be simplified to

$$2j\beta \frac{\partial \phi(x,z)}{\partial z} = \frac{\partial^2 \phi(x,z)}{\partial x^2}. \tag{5.146}$$

Introducing the difference parameter α, which determines the difference center in the z direction, we modify the wave equation (5.146) to get

$$2j\beta \frac{\phi(x, z+\Delta z) - \phi(z)}{\Delta z} = \alpha \frac{\partial^2 \phi(x, z+\Delta z)}{\partial x^2} + (1-\alpha)\frac{\partial^2 \phi(x,z)}{\partial x^2}. \tag{5.147}$$

Here, it should be noted that $\alpha = 0.5$ has been assumed in the preceding discussions and that the scheme with $\alpha = 0.5$ is called the Crank–Nicolson scheme.

When for a plane wave the slowly varying envelope function is expressed as

$$\phi(x,z) = \phi_0 \exp(jk_x x)\exp(-j\beta z), \tag{5.148}$$

the second derivative of the wave function with respect to x is

$$\frac{\partial^2 \phi(x,z)}{\partial x^2} = -k_x^2 \phi(x,z). \tag{5.149}$$

Substituting Eqs. (5.148) and (5.149) into (5.147) and dividing the resultant equation by $\phi(x, z)$, we get

$$2j\beta \frac{1}{\Delta z}[\exp(-j\beta \, \Delta z) - 1] = -k_x^2 \{\alpha \, \exp(-j\beta \, \Delta z) + (1 - \alpha)\}.$$

Therefore

$$\left(\frac{2j\beta}{\Delta z} + k_x^2 \alpha\right) \exp(-j\beta \, \Delta z) = \frac{2j\beta}{\Delta z} - (1 - \alpha)k_x^2$$

and

$$\exp(-j\beta \, \Delta z) = \frac{2j\beta/\Delta z - (1 - \alpha)k_x^2}{2j\beta/\Delta z + \alpha k_x^2}. \tag{5.150}$$

Since the exponential term $\exp(-j\beta \, \Delta z)$ given by Eq. (5.150) is a propagation term, we can clarify the influence of the parameter α on the stability of the beam propagation by investigating how the absolute value of the exponential function changes when the propagation distance Δz changes. The absolute value of the propagation term is expressed as

$$\lambda = |\exp(-j\beta \, \Delta z)|$$

$$= \left\{\frac{(2\beta/\Delta z)^2 + (1 - \alpha)^2 k_x^4}{(2\beta/\Delta z)^2 + \alpha^2 k_x^4}\right\}^{1/2}$$

$$= \left\{\frac{C^2 + (1 - \alpha)^2 k_x^4}{C^2 + \alpha^2 k_x^4}\right\}^{1/2}, \tag{5.151}$$

where

$$C = \frac{2\beta}{\Delta z}. \tag{5.152}$$

The absolute value of the propagation term is related to the value of α as follows:

CASE $\alpha = 0.5$

$$\lambda = \left\{\frac{C^2 + k_x^4/4}{C^2 + k_x^4/4}\right\}^{1/2} = 1. \tag{5.153}$$

Since the absolute value of the propagation term is always equal to 1, the propagating field does not diverge as the beam propagates. There is, however, the possibility of oscillation.

CASE $0.5 < \alpha \leq 1$

$$(1-\alpha)^2 < \alpha^2,$$

$$\therefore \lambda = \left\{\frac{C^2 + (1-\alpha)^2 k_x^4}{C^2 + \alpha^2 k_x^4}\right\}^{1/2} < 1. \quad (5.154)$$

Since the absolute value of the propagation term is always less than 1, the propagating field is unconditionally stable. However, the propagating field decays as the beam propagates.

CASE $0 \leq \alpha < 0.5$

$$(1-\alpha)^2 > \alpha^2,$$

$$\therefore \lambda = \left\{\frac{C^2 + (1-\alpha)^2 k_x^4}{C^2 + \alpha^2 k_x^4}\right\}^{1/2} > 1. \quad (5.155)$$

Since the absolute value of the propagation term is always greater than 1, the propagating field grows larger and larger as the beam propagates. It will finally diverge. A difference parameter α less than 0.5 thus should not be used, and $\alpha = 0.5$ is usually used in the calculation.

The Fresnel wave equation with the difference parameter α is expressed as

$$2j\beta \frac{\phi_p^{l+1} - \phi_p^l}{\Delta z} = \alpha(\alpha_w^l \phi_{p-1}^l + \{\alpha_x^l + k_0^2[\varepsilon_r^l(p) - n_{\text{eff}}^2]\}\phi_p^l + \alpha_e^l \phi_{p+1}^l)$$
$$+ (1-\alpha)(\alpha_w^{l+1} \phi_{p-1}^{l+1} + \{\alpha_x^{l+1} + k_0^2[\varepsilon_r^{l+1}(p) - n_{\text{eff}}^2]\}\phi_p^{l+1}$$
$$+ \alpha_e^{l+1} \phi_{p+1}^{l+1}). \quad (5.156)$$

Equation (5.156) is solved in exactly the same way that Eqs. (5.112) and (5.133), which are based on the Crank–Nicolson sheme, are solved.

5.2.5 Transparent Boundary Condition

An infinitely wide area would have no analysis boundary and there would thus be no reflections at boundaries. Because such an area cannot be assumed in actual design, however, a limited analysis window has to be used. In actual structures, radiated waves are reflected at the boundaries and return to the core area, where they interact with the propagating fields. This interaction disturbs the propagating fields and greatly degrades the calculation accuracy. In this section, we discuss the TBC. The TBC was developed by Hadley [5] as a way to efficiently suppress the reflections at boundaries, and it is easy to implement into computer programs. Although the TBC will be applied to 2D problems here, it can easily be extended and applied to 3D problems.

As shown in Fig. 5.6, the analysis window contains nodes at $p = 1$ to $p = M$. The hypothetical nodes at $p = 0$ and $p = M + 1$ are assumed to be outside the analysis window. Following Hadley's line of thinking [5], we incorporate the influences of the nodes at $p = 0$ and $p = M + 1$ into the nodes at $p = 1$ and $p = M$. In what follows, we consider how the boundary conditions can be written into a computer program.

A. Left-Hand Boundary Consider the left-hand boundary in Fig. 5.6. We incorporate the influence of the hypothetical node at $p = 0$ (outside the analysis window) into the node at $p = 1$ (inside the analysis window).

The wave function for the left-traveling wave with the x-directed wave number k_x is expressed as

$$\phi(x, z) = A(z) \exp(jk_x x). \tag{5.157}$$

We denote the x coordinates and the fields of the nodes at $p = 0, 1, 2$ as x_0, x_1, x_2 and as ϕ_0, ϕ_1, ϕ_2, and we assume that

$$\phi_0 = A(z) \exp(jk_x x_0), \tag{5.158}$$

$$\phi_1 = A(z) \exp(jk_x x_1), \tag{5.159}$$

$$\phi_2 = A(z) \exp(jk_x x_2). \tag{5.160}$$

The fields ϕ_1 and ϕ_2 are inside the analysis window, and ϕ_0 is a hypothetical field whose influence should be incorporated into the field inside the analysis window.

FIGURE 5.6. Nodes $p = 0$ and $p = M + 1$ are outside the analysis area.

Dividing Eq. (5.160) by Eq. (5.159) and Eq. (5.159) by Eq. (5.158), we get

$$\exp(jk_x \Delta x) = \frac{\phi_2}{\phi_1}, \quad (5.161)$$

$$\exp(jk_x \Delta x) = \frac{\phi_1}{\phi_0}, \quad (5.162)$$

where $\Delta x = x_2 - x_1 = x_1 - x_0$. Substituting the ratio of ϕ_2 to ϕ_1,

$$\eta_1 = \frac{\phi_2}{\phi_1}, \quad (5.163)$$

which is equal to $\exp(jk_x \Delta x)$, into Eq. (5.162), we get

$$\phi_0 = \frac{\phi_1}{\eta_1}. \quad (5.164)$$

Equation (5.164) can also be derived by substituting the x-directed wave number

$$k_x = \frac{1}{j \Delta x} \ln(\eta_1), \quad (5.165)$$

which is obtained from Eq. (5.161), into

$$\phi_0 = \phi_1 \exp(-jk_x \Delta x), \quad (5.166)$$

which is obtained from Eq. (5.162).

It should be noted that since the wave travels leftward, the real part of the x-directed wave number k_x, $\text{Re}(k_x)$, should be negative. When it is positive, which implies reflection at the left-hand boundary, the sign should be changed from plus to minus.

B. Right-Hand Boundary Consider the right-hand boundary in Fig. 5.6. We incorporate the influence of the hypothetical node at $p = M + 1$ (outside the analysis window) into the node at $p = M$ (inside the analysis window).

BEAM PROPAGATION METHODS

The wave function of the right-traveling wave with the x-directed wave number k_x is expressed as

$$\phi(x, z) = A(z) \exp(-jk_x x). \tag{5.167}$$

We denote the x coordinates and the fields of the nodes $p = M - 1, M, M + 1$ as x_{M-1}, x_M, x_{M+1} and $\phi_{M-1}, \phi_M, \phi_{M+1}$, and we assume that

$$\phi_{M-1} = A(z) \exp(-jk_x x_{M-1}), \tag{5.168}$$

$$\phi_M = A(z) \exp(-jk_x x_M), \tag{5.169}$$

$$\phi_{M+1} = A(z) \exp(-jk_x x_{M+1}), \tag{5.170}$$

where ϕ_{M-1} and ϕ_M are the fields inside the analysis window and ϕ_{M+1} is the hypothetical field whose influence should be incorporated into the field inside the analysis window.

Dividing Eq. (5.169) by Eq. (5.168) and Eq. (5.170) by Eq. (5.169), we get

$$\exp(-jk_x \Delta x) = \frac{\phi_{M+1}}{\phi_M} \tag{5.171}$$

$$\exp(-jk_x \Delta x) = \frac{\phi_M}{\phi_{M-1}}, \tag{5.172}$$

where $\Delta x = x_M - x_{M-1} = x_{M+1} - x_M$. Substituting the ratio of ϕ_M to ϕ_{M-1},

$$\eta_M = \frac{\phi_M}{\phi_{M-1}}, \tag{5.173}$$

which is equal to $\exp(jk_x \Delta x)$, into Eq. (5.172), we get

$$\phi_{M+1} = \phi_M \eta_M. \tag{5.174}$$

Equation (5.174) can also be derived by substituting the x-directed wave number

$$k_x = \frac{1}{j \Delta x} \ln(\eta_M), \tag{5.175}$$

which is obtained from Eq. (5.171), into

$$\phi_{M+1} = \phi_M \exp(-jk_x \Delta x). \tag{5.176}$$

Similar to what we saw in the case of the left-hand boundary, since the wave travels rightward, the real part of the x-directed wave number k_x should be negative. When it is positive, which implies reflection occurs at the right-hand boundary, the sign should be changed from plus to minus.

5.2.6 Programming

We now look at how the TBC is written into the program. For the FD scheme in the propagation direction, the most widely used Crank–Nicolson scheme (i.e., $\alpha = 0.5$) in Eq. (5.147) is used.

The problem is to obtain the unknown field ϕ^{l+1} at $z + \Delta z$ by using the known field ϕ^l, where superscripts l and $l+1$ respectively correspond to z and $z + \Delta z$. The following equation has to be solved:

$$-\alpha_w^{l+1} \phi_{p-1}^{l+1} + \left\{ -\alpha_x^{l+1} + \frac{4j\beta}{\Delta z} - k_0^2 [\varepsilon_r^{l+1}(p) - n_{\text{eff}}^2] \right\} \phi_p^{l+1} - \alpha_e^{l+1} \phi_{p+1}^{l+1}$$
$$= \alpha_w^l \phi_{p-1}^l + \left\{ \alpha_x^l + \frac{4j\beta}{\Delta z} + k_0^2 [\varepsilon_r^l(p) - n_{\text{eff}}^2] \right\} \phi_p^l + \alpha_e^l \phi_{p+1}^l. \tag{5.177}$$

The unknown fields to be obtained at $z + \Delta z$ are ϕ_{p-1}^{l+1}, ϕ_p^{l+1}, and ϕ_{p+1}^{l+1} and the known fields at z are ϕ_{p-1}^l, ϕ_p^l, and ϕ_{p+1}^l for $p = 1$ and M. Assuming the unknown coefficients $A(p)$, $B(p)$, and $C(p)$ and the known value $D(p)$ to be

$$A(p) = -\alpha_w^{l+1}, \tag{5.178}$$

$$B(p) = \left\{ -\alpha_x^{l+1} + \frac{4j\beta}{\Delta z} - k_0^2 [\varepsilon_r^{l+1}(p) - n_{\text{eff}}^2] \right\}, \tag{5.179}$$

$$C(p) = -\alpha_e^{l+1}, \tag{5.180}$$

$$D(p) = \alpha_w^l \phi_{p-1}^l + \left\{ \alpha_x^l + \frac{4j\beta}{\Delta z} + k_0^2 [\varepsilon_r^l(p) - n_{\text{eff}}^2] \right\} \phi_p^l + \alpha_e^l \phi_{p+1}^l, \tag{5.181}$$

we simplify Eq. (5.177) to

$$A(p)\phi_{p-1}^{l+1} + B(p)\phi_p^{l+1} + C(p)\phi_{p+1}^{l+1} = D(p). \tag{5.182}$$

In the following, we will discuss Eq. (5.182) with respect to p, which represents the lateral position of the node.

A. Left-Hand Boundary ($p = 1$) The field ϕ_1 of the node on the left-hand boundary is influenced by the field ϕ_0 of the hypothetical node outside the analysis window. As shown in Eq. (5.164), the field ϕ_0 of the hypothetical node is expressed as

$$\phi_0 = \phi_1 \gamma_L, \tag{5.183}$$

where

$$\gamma_L = \frac{1}{\eta_1} = \frac{1}{\phi_2^l / \phi_1^l}. \tag{5.184}$$

Here, since the parameter γ_L is determined by the known fields at z (i.e., l), γ_L is known. Thus, Eq. (5.182) is reduced to

$$B'(1)\phi_1^{l+1} + C(1)\phi_2^{l+1} = D(1), \tag{5.185}$$

where

$$B'(1) = -\alpha_w^{l+1} \gamma_L + \left\{ -\alpha_x^{l+1} + \frac{4j\beta}{\Delta z} - k_0^2 [\varepsilon_r^{l+1}(1) - n_{\text{eff}}^2] \right\}, \tag{5.186}$$

$$C(1) = -\alpha_e^{l+1}, \tag{5.187}$$

$$D(1) = \alpha_w^l \gamma_L \phi_1^l + \left\{ \alpha_x^l + \frac{4j\beta}{\Delta z} + k_0^2 [\varepsilon_r^l(1) - n_{\text{eff}}^2] \right\} \phi_1^l + \alpha_e^l \phi_2^l. \tag{5.188}$$

B. Right-Hand Boundary ($p = M$) The field ϕ_M of the node on the right-hand boundary is influenced by the field ϕ_{M+1} of the hypothetical node outside the analysis window. As shown in Eq. (5.174), the field ϕ_{M+1} of the hypothetical node is expressed as

$$\phi_{M+1} = \phi_M \gamma_R, \tag{5.189}$$

where

$$\gamma_R = \eta_M = \frac{1}{\phi_M^l / \phi_{M-1}^l}. \tag{5.190}$$

Here, since the parameter γ_R is determined by the known fields at z (i.e., l), γ_L is known. Thus, Eq. (5.182) is reduced to

$$A(M)\phi_{M-1}^{l+1} + B'(M)\phi_M^{l+1} = D(M), \qquad (5.191)$$

where

$$A(M) = -\alpha_w^{l+1}, \qquad (5.192)$$

$$B'(M) = -\alpha_e^{l+1}\gamma_R + \left\{-\alpha_x^{l+1} + \frac{4j\beta}{\Delta z} - k_0^2[\varepsilon_r^{l+1}(M) - n_{\text{eff}}^2]\right\}, \qquad (5.193)$$

$$D(M) = \alpha_w^l \phi_{M-1}^l + \left\{\alpha_e^l \gamma_R + \alpha_x^l + \frac{4j\beta}{\Delta z} + k_0^2[\varepsilon_r^l(M) - n_{\text{eff}}^2]\right\}\phi_M^l. \qquad (5.194)$$

Summarizing the above discussions, we get the algebraic equations

$$\begin{pmatrix} B'(1) & C(1) & & & & \\ A(2) & B(2) & C(2) & & & \\ & A(3) & B(3) & C(3) & & \\ & & \ddots & \ddots & \ddots & \\ & & & & & C(M-1) \\ & & & & A(M) & B'(M) \end{pmatrix} \begin{pmatrix} \phi_1^{l+1} \\ \phi_2^{l+1} \\ \phi_3^{l+1} \\ \vdots \\ \phi_{M-1}^{l+1} \\ \phi_M^{l+1} \end{pmatrix}$$

$$= \begin{pmatrix} D(1) \\ D(2) \\ D(3) \\ \vdots \\ D(M-1) \\ D(M) \end{pmatrix}. \qquad (5.195)$$

Since Eq. (5.195) is a tridiagonal matrix equation, we can obtain the unknown wave fields $\phi_1^{l+1}, \ldots, \phi_M^{l+1}$ by using the Thomas method [14].

5.3 WIDE-ANGLE ANALYSIS USING PADÉ APPROXIMANT OPERATORS

Up to this stage, the discussions have been based on the Fresnel equation (i.e., the para-axial wave equation). Here, we discuss the wide-angle beam propagation method based on Padé approximant operators [5] and discuss the multistep method [6], both of which were developed by Hadley.

5.3.1 Padé Approximant Operators

When the second derivative with respect to z is not neglected, the wave equation is

$$2j\beta \frac{\partial \phi}{\partial z} - \frac{\partial^2 \phi}{\partial z^2} = P\phi, \tag{5.196}$$

where for the TE mode

$$P = \frac{\partial^2}{\partial x^2} + k_0^2(\varepsilon_r - n_{\text{eff}}^2) \tag{5.197}$$

and for the TM mode

$$P = \varepsilon_r \frac{\partial}{\partial x}\left(\frac{1}{\varepsilon_r}\frac{\partial}{\partial x}\right) + k_0^2(\varepsilon_r - n_{\text{eff}}^2). \tag{5.198}$$

Solving Eq. (5.196) formally, we get

$$\frac{\partial}{\partial z}\left(1 + \frac{j}{2\beta}\frac{\partial}{\partial z}\right)\phi = -\frac{jP}{2\beta}\phi$$

and therefore

$$\frac{\partial \phi}{\partial z} = \frac{-jP/2\beta}{1 + (j/2\beta)(\partial/\partial z)}\phi. \tag{5.199}$$

5.3 WIDE-ANGLE ANALYSIS USING PADÉ APPROXIMANT OPERATORS

When the derivative with respect to z is neglected, Eq. (5.199) is reduced to the Fresnel equation. Here, we regard the derivative with respect to z in Eq. (5.199) as the recurrence formula

$$\left.\frac{\partial}{\partial z}\right|_n = \left.\frac{-jP/2\beta}{1+(j/2\beta)(\partial/\partial z)}\right|_{n-1}. \tag{5.200}$$

Next, we will specify the explicit expressions for the various orders of the recurrence formula. First, we define the starting equation

$$\left.\frac{\partial}{\partial z}\right|_{-1} = 0. \tag{5.201}$$

The explicit expressions for the corresponding wide-angle (WA) orders are shown below.

1. WA-0th order (Fresnel approximation):

$$\left.\frac{\partial}{\partial z}\right|_0 = \frac{-jP/2\beta}{1+\dfrac{j}{2\beta}\left.\dfrac{\partial}{\partial z}\right|_{-1}} = -j\frac{P}{2\beta}. \tag{5.202}$$

2. WA-1st order:

$$\left.\frac{\partial}{\partial z}\right|_1 = \frac{-jP/2\beta}{1+\dfrac{j}{2\beta}\left.\dfrac{\partial}{\partial z}\right|_0} = \frac{-jP/2\beta}{1+\dfrac{j}{2\beta}\left(-\dfrac{jP}{2\beta}\right)} = -j\frac{P/2\beta}{1+P/4\beta^2}. \tag{5.203}$$

3. WA-2nd order:

$$\left.\frac{\partial}{\partial z}\right|_2 = \frac{-jP/2\beta}{1+\dfrac{j}{2\beta}\left.\dfrac{\partial}{\partial z}\right|_1} = \frac{-jP/2\beta}{1+\dfrac{j}{2\beta}\dfrac{-jP/2\beta}{1+P/4\beta^2}}$$

$$= -j\frac{P/2\beta + P^2/8\beta^3}{1+P/2\beta^2}. \tag{5.204}$$

206 BEAM PROPAGATION METHODS

4. WA-3rd order:

$$\left.\frac{\partial}{\partial z}\right|_3 = \frac{-jP/2\beta}{1+\dfrac{j}{2\beta}\left.\dfrac{\partial}{\partial z}\right|_2} = \frac{-jP/2\beta}{1+\dfrac{j}{2\beta}\dfrac{-j(P/2\beta+P^2/8\beta^3)}{1+P/2\beta^2}}$$

$$= \frac{-(jP/2\beta)(1+P/2\beta^2)}{1+P/2\beta^2+P/4\beta^2+P^2/16\beta^4}$$

$$= -j\frac{P/2\beta+P^2/4\beta^3}{1+3P/4\beta^2+P^2/16\beta^4}. \tag{5.205}$$

5. WA-4th order:

$$\left.\frac{\partial}{\partial z}\right|_4 = \frac{-jP/2\beta}{1+\dfrac{j}{2\beta}\left.\dfrac{\partial}{\partial z}\right|_3} = \frac{-jP/2\beta}{1+\dfrac{j}{2\beta}\dfrac{-j(P/2\beta+P^2/4\beta^3)}{1+3P/4\beta^2+P^2/16\beta^4}}$$

$$= \frac{-(jP/2\beta)(1+3P/4\beta^2+P^2/16\beta^4)}{1+3P/4\beta^2+P^2/16\beta^4+P/4\beta^2+P^2/8\beta^4}$$

$$= -j\frac{P/2\beta+3P^2/8\beta^3+P^3/32\beta^5}{1+P/\beta^2+3P^2/16\beta^4}. \tag{5.206}$$

6. WA-5th order:

$$\left.\frac{\partial}{\partial z}\right|_5 = \frac{-jP/2\beta}{1+\dfrac{j}{2\beta}\left.\dfrac{\partial}{\partial z}\right|_4} = \frac{-jP/2\beta}{1+\dfrac{j}{2\beta}\dfrac{-j(P/2\beta+3P^2/8\beta^3+P^3/32\beta^5)}{1+P/\beta^2+3P^2/16\beta^4}}$$

$$= \frac{-(jP/2\beta)(1+P/\beta^2+3P^2/16\beta^4)}{1+P/\beta^2+3P^2/16\beta^4+P/4\beta^2+3P^2/16\beta^4+P^3/64\beta^6}$$

$$= -j\frac{P/2\beta+P^2/2\beta^3+3P^3/32\beta^5}{1+5P/4\beta^2+3P^2/8\beta^4+P^3/64\beta^6}. \tag{5.207}$$

5.3 WIDE-ANGLE ANALYSIS USING PADÉ APPROXIMANT OPERATORS

7. WA-6th order:

$$\left.\frac{\partial}{\partial z}\right|_6 = \frac{-jP/2\beta}{1+\dfrac{j}{2\beta}\left.\dfrac{\partial}{\partial z}\right|_5} = \frac{-jP/2\beta}{1+\dfrac{j}{2\beta}\dfrac{-j(P/2\beta + P^2/2\beta^3 + 3P^3/32\beta^5)}{1+5P/4\beta^2 + 3P^2/8\beta^4 + P^3/64\beta^6}}$$

$$= \frac{-(jP/2\beta)(1+5P/4\beta^2 + 3P^2/8\beta^4 + P^3/64\beta^6)}{\begin{array}{c}1+5P/4\beta^2 + 3P^2/8\beta^4 + P^3/64\beta^6\\+P/4\beta^2 + P^2/4\beta^4 + 3P^3/64\beta^6\end{array}}$$

$$= \frac{-j(P/2\beta + 5P^2/8\beta^3 + 3P^3/16\beta^5 + P^4/128\beta^7)}{1+3P/2\beta^2 + 5P^2/8\beta^4 + P^3/16\beta^6}$$

$$= -j\frac{P/2\beta + 5P^2/8\beta^3 + 3P^3/16\beta^5 + P^4/128\beta^7}{1+3P/2\beta^2 + 5P^2/8\beta^4 + P^3/16\beta^6}. \tag{5.208}$$

8. WA-7th order:

$$\left.\frac{\partial}{\partial z}\right|_7 = \frac{-jP/2\beta}{1+\dfrac{j}{2\beta}\left.\dfrac{\partial}{\partial z}\right|_6}$$

$$= \frac{-jP/2\beta}{1+\dfrac{j}{2\beta}\dfrac{-j(P/2\beta + 5P^2/8\beta^3 + 3P^3/16\beta^5 + P^4/128\beta^7)}{1+3P/2\beta^2 + 5P^2/8\beta^4 + P^3/16\beta^6}}$$

$$= \frac{-(jP/2\beta)(1+3P/2\beta^2 + 5P^2/8\beta^4 + P^3/16\beta^6)}{\begin{array}{c}1+3P/2\beta^2 + 5P^2/8\beta^4 + P^3/16\beta^6 + P/4\beta^2\\+5P^2/16\beta^4 + 3P^2/32\beta^6 + P^4/256\beta^8\end{array}}$$

$$= -j\frac{P/2\beta + 3P^2/4\beta^3 + 5P^3/16\beta^5 + P^4/32\beta^7}{1+7P/4\beta^2 + 15P^2/16\beta^4 + 5P^3/32\beta^6 + P^4/256\beta^8}. \tag{5.209}$$

Thus, the recurrence formula (5.200) can be reduced to an expression that includes only the operator P:

$$\frac{\partial \phi}{\partial z} = -j\frac{N}{D}\phi, \tag{5.210}$$

where N and D are both polynomials of the operator P. The wide-angle orders correspond to the orders for the Padé orders as follows:

$$WA - 0 \leftrightarrow (1, 0),$$
$$WA - 1 \leftrightarrow (1, 1),$$
$$WA - 2 \leftrightarrow (2, 1),$$
$$WA - 3 \leftrightarrow (2, 2),$$
$$WA - 4 \leftrightarrow (3, 2),$$
$$WA - 5 \leftrightarrow (3, 3),$$
$$WA - 6 \leftrightarrow (4, 3),$$
$$WA - 7 \leftrightarrow (4, 4).$$

Differentiating Eq. (5.210) based on the Crank–Nicolson scheme, we get

$$\text{Left-hand side:} \quad \frac{\partial \phi}{\partial z} \rightarrow \frac{1}{\Delta z}(\phi^{l+1} - \phi^l),$$
$$\text{Right-hand side:} \quad -j\frac{N}{D}\phi \rightarrow -j\frac{N}{D}\frac{1}{2}(\phi^{l+1} + \phi^l), \quad (5.211)$$

where the right-hand side was averaged by l and $l+1$ so that the difference center of the left-hand side coincides with that of the right-hand side.

From Eq. (5.211), we get

$$\frac{1}{\Delta z}(\phi^{l+1} - \phi^l) = -j\frac{N}{D}\frac{1}{2}(\phi^{l+1} + \phi^l).$$

Therefore

$$D(\phi^{l+1} - \phi^l) = -jN\,\Delta z = \frac{1}{2}(\phi^{l+1} + \phi^l)$$

and

$$\left(D + j\frac{\Delta z}{2}N\right)\phi^{l+1} = \left(D - j\frac{\Delta z}{2}N\right)\phi^l, \quad (5.212)$$

which can be rewritten as

$$\phi^{l+1} = \frac{D - j(\Delta z/2)N}{D + j(\Delta z/2)N}\phi^l. \quad (5.213)$$

Since, as shown in Eqs. (5.202)–(5.209), the coefficients of the polynomials D and N in Eq. (5.213) are real, D and N themselves are real. Thus, Eq. (5.213) can be written as

$$\phi^{l+1} = \frac{D - j(\Delta z/2)N}{[D - j(\Delta z/2)N]^*} \phi^l$$
$$= \frac{\sum_{i=0}^{n} \xi_i P^i}{\sum_{i=0}^{n} \xi_i^* P^i} \phi^l. \tag{5.214}$$

In the following, we show the coefficients ξ_i for the wide-angle orders WA-0 to WA-7.

1. WA-0th order [Padé(1,0): Fresnel approximation]: From Eq. (5.202), we get

$$D = 1, \qquad N = \frac{P}{2\beta}$$

and therefore

$$D - j\frac{\Delta z}{2} N = 1 - j\frac{\Delta z}{2} \frac{P}{2\beta}. \tag{5.215}$$

Thus, we get

$$\xi_0 = 1, \qquad \xi_1 = -j\frac{\Delta z}{4\beta}. \tag{5.216}$$

2. WA-1st order [Padé(1,1)]: From Eq. (5.203), we get

$$D = 1 + \frac{P}{4\beta^2}, \qquad N = \frac{P}{2\beta}$$

and therefore

$$D - j\frac{\Delta z}{2} N = 1 + \frac{P}{4\beta^2} - j\Delta \frac{P}{4\beta} = 1 + \frac{1}{4\beta^2}(1 - j\beta \, \Delta z)P. \tag{5.217}$$

Thus, we get

$$\xi_0 = 1, \qquad \xi_1 = \frac{1}{4\beta^2}(1 - j\beta \Delta z). \qquad (5.218)$$

3. WA-2nd order [Padé(2,1)]: From Eq. (5.204), we get

$$D = 1 + \frac{P}{2\beta^2}, \qquad N = \frac{P}{2\beta} + \frac{P^2}{8\beta^3}$$

and therefore

$$D - j\frac{\Delta z}{2} N = 1 + \frac{P}{2\beta^2} - j\frac{\Delta z}{2}\left(\frac{P}{2\beta} + \frac{P^2}{8\beta^3}\right)$$
$$= 1 + \frac{1}{4\beta^2}(2 - j\beta \Delta z)P - j\frac{\Delta z}{16\beta^3}P^2. \qquad (5.219)$$

Thus, we get

$$\xi_0 = 1, \qquad \xi_1 = \frac{1}{4\beta^2}(2 - j\beta \Delta z), \qquad \xi_2 = -j\frac{\Delta z}{16\beta^3}. \qquad (5.220)$$

4. WA-3rd order [Padé(2,2)]: From Eq. (5.205), we get

$$D = 1 + \frac{3P}{4\beta^2} + \frac{P^2}{16\beta^4}, \qquad N = \frac{P}{2\beta} + \frac{P^2}{4\beta^3}$$

and therefore

$$D - j\frac{\Delta z}{2} N = 1 + \frac{3P}{4\beta^2} + \frac{P^2}{16\beta^4} - j\frac{\Delta z}{2}\left(\frac{P}{2\beta} + \frac{P^2}{4\beta^3}\right)$$
$$= 1 + \frac{1}{4\beta^2}(3 - j\beta \Delta z)P + \frac{1}{16\beta^4}(1 - j2\beta \Delta z)P^2. \qquad (5.221)$$

Thus, we get

$$\xi_0 = 1, \qquad \xi_1 = \frac{1}{4\beta^2}(3 - j\beta \Delta z), \qquad \xi_2 = \frac{1}{16\beta^4}(1 - j2\beta \Delta z). \qquad (5.222)$$

5. **WA-4th order [Padé(3,2)]**: From Eq. (5.206), we get

$$D = 1 + \frac{P}{\beta^2} + \frac{3P^2}{16\beta^4}, \qquad N = \frac{P}{2\beta} + \frac{3P^2}{8\beta^3} + \frac{P^3}{32\beta^5}$$

and therefore

$$D - j\frac{\Delta z}{2}N = 1 + \frac{P}{\beta^2} + \frac{3P^2}{16\beta^4} - j\frac{\Delta z}{2}\left(\frac{P}{2\beta} + \frac{3P^2}{8\beta^3} + \frac{P^3}{32\beta^5}\right)$$

$$= 1 + \frac{1}{4\beta^2}(4 - j\beta\,\Delta z)P + \frac{1}{16\beta^4}(3 - j3\beta\,\Delta z)P^2$$

$$- j\frac{\Delta z}{64\beta^5}P^3. \qquad (5.223)$$

Thus, we get

$$\xi_0 = 1, \qquad \xi_1 = \frac{1}{4\beta^2}(4 - j\beta\,\Delta z), \qquad \xi_2 = \frac{1}{16\beta^4}(3 - j3\beta\,\Delta z),$$

$$\xi_3 = -j\frac{\Delta z}{64\beta^5}. \qquad (5.224)$$

6. **WA-5th order [Padé(3,3)]**: From Eq. (5.207), we get

$$D = 1 + \frac{5P}{4\beta^2} + \frac{3P^2}{8\beta^4} + \frac{P^3}{64\beta^6}, \qquad N = \frac{P}{2\beta} + \frac{P^2}{2\beta^3} + \frac{3P^3}{32\beta^5},$$

and therefore

$$D - j\frac{\Delta z}{2}N = 1 + \frac{5P}{4\beta^2} + \frac{3P^2}{8\beta^4} + \frac{P^3}{64\beta^6} - j\frac{\Delta z}{2}\left(\frac{P}{2\beta} + \frac{P^2}{2\beta^3} + \frac{3P^2}{32\beta^5}\right)$$

$$= 1 + \frac{1}{4\beta^2}(5 - j\beta\,\Delta z)P + \frac{1}{8\beta^4}(3 - j2\beta\,\Delta z)P^2$$

$$+ \frac{1}{64\beta^6}(1 - j3\beta\,\Delta z)P^3. \qquad (5.225)$$

Thus, we get

$$\xi_0 = 1, \quad \xi_1 = \frac{1}{4\beta^2}(5 - j\beta\,\Delta z), \quad \xi_2 = \frac{1}{8\beta^4}(3 - j2\beta\,\Delta z),$$

$$\xi_3 = \frac{1}{64\beta^6}(1 - j3\beta\,\Delta z). \tag{5.226}$$

7. WA-6th order [Padé(4,3)]: From Eq. (5.208), we get

$$D = 1 + \frac{3P}{2\beta^2} + \frac{5P^2}{8\beta^4} + \frac{P^3}{16\beta^6},$$

$$N = \frac{P}{2\beta} + \frac{5P^2}{8\beta^3} + \frac{3P^3}{16\beta^5} + \frac{P^4}{128\beta^7},$$

and therefore

$$\begin{aligned} D - j\frac{\Delta z}{2}N &= 1 + \frac{3P}{2\beta^2} + \frac{5P^2}{8\beta^4} + \frac{P^3}{16\beta^6} \\ &\quad - j\frac{\Delta z}{2}\left(\frac{P}{2\beta} + \frac{5P^2}{8\beta^3} + \frac{3P^3}{16\beta^5} + \frac{P^4}{128\beta^7}\right) \\ &= 1 + \frac{1}{4\beta^2}(6 - j\beta\,\Delta z)P + \frac{1}{16\beta^4}(10 - j5\beta\,\Delta z)P^2 \\ &\quad + \frac{1}{32\beta^6}(2 - j3\beta\,\Delta z)P^3 - \frac{j\,\Delta z}{256\beta^7}P^4. \end{aligned} \tag{5.227}$$

Thus, we get

$$\xi_0 = 1, \quad \xi_1 = \frac{1}{4\beta^2}(6 - j\beta\,\Delta z), \quad \xi_2 = \frac{1}{16\beta^4}(10 - j5\beta\,\Delta z),$$

$$\xi_3 = \frac{1}{32\beta^6}(2 - j3\beta\,\Delta z), \quad \xi_4 = -\frac{j\,\Delta z}{256\beta^7}. \tag{5.228}$$

8. WA-7th order [Padé(4,4)]: From Eq. (5.209), we get

$$D = 1 + \frac{7P}{4\beta^2} + \frac{15P^2}{16\beta^4} + \frac{5P^3}{32\beta^6} + \frac{P^4}{256\beta^8},$$

$$N = \frac{P}{2\beta} + \frac{3P^2}{4\beta^3} + \frac{5P^3}{16\beta^5} + \frac{P^4}{32\beta^7},$$

and therefore

$$\begin{aligned}D - j\frac{\Delta z}{2}N &= 1 + \frac{7P}{4\beta^2} + \frac{15P^2}{16\beta^4} + \frac{5P^3}{32\beta^6} + \frac{P^4}{256\beta^8} \\ &\quad - j\frac{\Delta z}{2}\left(\frac{P}{2\beta} + \frac{3P^2}{4\beta^3} + \frac{5P^3}{16\beta^5} + \frac{P^4}{32\beta^7}\right) \\ &= 1 + \frac{1}{4\beta^2}(7 - j\beta\,\Delta z)P + \frac{1}{16\beta^4}(15 - j6\beta\,\Delta z)P^2 \\ &\quad + \frac{1}{32\beta^6}(5 - j5\beta\,\Delta z)P^3 + \frac{1}{256\beta^8}(1 - j4\beta\,\Delta z)P^4.\end{aligned}$$
(5.229)

Thus, we get

$$\xi_0 = 1, \quad \xi_1 = \frac{1}{4\beta^2}(7 - j\beta\,\Delta z), \quad \xi_2 = \frac{1}{16\beta^4}(15 - j6\beta\,\Delta z),$$

$$\xi_3 = \frac{1}{32\beta^6}(5 - j5\beta\,\Delta z), \quad \xi_4 = \frac{1}{256\beta^8}(1 - j4\beta\,\Delta z). \quad (5.230)$$

Since through the above discussions the explicit expressions for $D - j(\Delta z/2)N$ are obtained, those for $D + j(\Delta z/2)N$ can also be obtained. Both sides of Eq. (5.212) are clarified and the unknown field ϕ^{l+1} can be calculated.

5.3.2 Multistep Method

The reason the matrix of the Fresnel approximation is tridiagonal is that the order of the operator P, which contains the second derivative with respect to the x coordinate and can be approximated by the FD scheme with three terms as shown in Eq. (4.9), is 1.

Since the wide-angle formulations written in Eqs. (5.204)–(5.209) or Eqs. (5.219)–(5.230) include the powers of the operator P higher than 2, the column width of nonzero matrix elements is greater than 3. Thus, the numerically efficient Thomas method cannot be used to solve the final algebraic matrix equation. In this section, we discuss the multistep method, which was developed by Hadley [6] in order to solve this problem.

Consider Eq. (5.214). The numerator of the factor on the right-hand side of Eq. (5.214) is obtained as shown in Eqs. (5.215)–(5.230). The denominator can also be obtained, since it is simply the complex conjugate of the numerator. Since ξ_0 is equal to 1, the numerator of the term on the right-hand side of Eq. (5.214) can be factorized as

$$\sum_{i=0}^{n} \xi_i P^i = (1 + a_n P) \cdots (1 + a_2 P)(1 + a_1 P), \quad (5.231)$$

where the coefficients $a_1, a_2, \ldots, a_n$ can be obtained by solving the algebraic equation

$$D - j\frac{\Delta z}{2} N = \sum_{i=0}^{n} \xi_i P^i = 0. \quad (5.232)$$

The denominator of the term on the right-hand side of Eq. (5.214) can be obtained by using the complex conjugates of the coefficients $a_1, a_2, \ldots, a_n$:

$$\sum_{i=0}^{n} \xi_i^* P^i = (1 + a_n P)^* \cdots (1 + a_2 P)^*(1 + a_1 P)^*$$
$$= (1 + a_n^* P) \cdots (1 + a_2^* P)(1 + a_1^* P), \quad (5.233)$$

Thus, the unknown field ϕ^{l+1} at $z + \Delta z$ is related to the known field ϕ^l at z as follows:

$$\phi^{l+1} = \frac{(1 + a_n P) \cdots (1 + a_2 P)(1 + a_1 P)}{(1 + a_n^* P) \cdots (1 + a_2^* P)(1 + a_1^* P)} \phi^l. \quad (5.234)$$

Next, we discuss how to solve Eq. (5.234). First, we rewrite it as

$$\frac{(1 + a_n^* P) \cdots (1 + a_2^* P)}{(1 + a_n P) \cdots (1 + a_2 P)} \phi^{l+1} = \frac{1 + a_1 P}{1 + a_1^* P} \phi^l. \quad (5.235)$$

5.3 WIDE-ANGLE ANALYSIS USING PADÉ APPROXIMANT OPERATORS

Then, defining the field $\phi^{l+1/n}$ as

$$\phi^{l+1/n} = \frac{(1+a_n^*P)\cdots(1+a_2^*P)}{(1+a_nP)\cdots(1+a_2P)}\phi^{l+1}, \tag{5.236}$$

we rewrite Eq. (5.235) as

$$\phi^{l+1/n} = \frac{1+a_1P}{1+a_1^*P}\phi^l. \tag{5.237}$$

Since ϕ^l is known, we can obtain $\phi^{l+1/n}$ by solving Eq. (5.237). Using $\phi^{l+1/n}$, we rewrite Eq. (5.236) as

$$\frac{(1+a_n^*P)\cdots(1+a_3^*P)}{(1+a_nP)\cdots(1+a_3P)}\phi^{l+1} = \frac{1+a_2P}{1+a_2^*P}\phi^{l+1/n}. \tag{5.238}$$

Then, defining the field $\phi^{l+2/n}$ as

$$\phi^{l+2/n} = \frac{(1+a_n^*P)\cdots(1+a_3^*P)}{(1+a_nP)\cdots(1+a_3P)}\phi^{l+1}, \tag{5.239}$$

we rewrite Eq. (5.238) as

$$\phi^{l+2/n} = \frac{1+a_2P}{1+a_2^*P}\phi^{l+1/n}. \tag{5.240}$$

Since $\phi^{l+1/n}$ is known, we can obtain $\phi^{l+2/n}$ by solving Eq. (5.240). Repeating this procedure, we finally get the unknown field ϕ^{l+1} at $z+\Delta z$ by solving

$$\phi^{l+1} = \frac{1+a_nP}{1+a_n^*P}\phi^{l+(n-1)/n}. \tag{5.241}$$

That is, the known field ϕ^{l+1} can be obtained from the known field ϕ^l by successively solving

$$\phi^{l+i/n} = \frac{1+a_iP}{1+a_i^*P}\phi^{l+(i-1)/n} \tag{5.242}$$

when $i = 1, 2, \ldots, n$.

The advantage of the multistep method is that the matrix equation to be solved in each step is the same size as the Fresnel equation and for 2D problems is tridiagonal. Thus, the calculation procedure is very easy. The method can be easily extended to 3D problems, and it has also been used in the wide-angle FE-BPM [12].

5.4 THREE-DIMENSIONAL SEMIVECTORIAL ANALYSIS

The preceding discussions were limited to the 2D BPM, which assumes the 1D cross-sectional structure in the lateral direction. When, however, the propagating field is widely spread in the 2D cross section, the 3D beam propagation method is required. Since the formulation for the 2D BPM in Section 5.2.2 can be straightforwardly extended to the 3D BPM, a much more numerically efficient 3D BPM formulation based on the alternate-direction implicit (ADI) method [8–10] will be shown here.

In the ADI-BPM, the calculation for one step, $z \to z + \Delta z$, is divided into two steps, $z \to z + \Delta z/2$ and $z + \Delta z/2 \to z + \Delta z$, and the two steps are solved successively in the x and y directions. Since solving a 3D problem can be reduced to solving a 2D problem twice by using the ADI method, instead of a large matrix equation we have only to solve tridiagonal matrix equations twice. Thus, high numerical efficiency is attained especially for large-size waveguides, such as spotsize-converter-integrated structures [15–17]. In this section, the semivectorial formulation is used to analyze large-index-difference waveguides and to treat the polarization. Nonuniform discretization is also assumed for versatility of analysis.

Neglecting the terms for the interaction between polarizations in the vectorial wave equations (4.10) and (4.19) in Section 4.2, we get the semivectorial wave equation

$$\frac{\partial^2 \psi}{\partial z^2} + P\psi = 0. \tag{5.243}$$

Here, $P\psi$ and ψ for the quasi-TE mode are obtained from Eqs. (4.17) and (4.30) as

$$P\psi = \frac{\partial}{\partial x}\left\{\frac{1}{\varepsilon_r}\frac{\partial}{\partial x}(\varepsilon_r \psi)\right\} + \frac{\partial^2 \psi}{\partial y^2} + k_0^2 \varepsilon_r \psi \quad \text{(electric field representation)},$$

$$\psi = E_x, \tag{5.244}$$

$$P\psi = \varepsilon_r \frac{\partial}{\partial x}\left(\frac{1}{\varepsilon_r}\frac{\partial \psi}{\partial x}\right) + \frac{\partial^2 \psi}{\partial y^2} + k_0^2 \varepsilon_r \psi \quad \text{(magnetic field representation)},$$

$$\psi = H_y. \tag{5.245}$$

5.4 THREE-DIMENSIONAL SEMIVECTORIAL ANALYSIS

And $P\psi$ and ψ for the quasi-TM mode are obtained from Eqs. (4.18) and (4.29) as

$$P\psi = \frac{\partial^2 \psi}{\partial x^2} + \frac{\partial}{\partial y}\left\{\frac{1}{\varepsilon_r}\frac{\partial}{\partial y}(\varepsilon_r \psi)\right\} + k_0^2 \varepsilon_r \psi \text{ (electric field representation)},$$

$$\psi = E_y, \qquad (5.246)$$

$$P\psi = \frac{\partial^2 \psi}{\partial x^2} + \varepsilon_r \frac{\partial}{\partial y}\left(\frac{1}{\varepsilon_r}\frac{\partial \psi}{\partial y}\right) + \frac{\partial^2 \psi}{\partial z^2} + k_0^2 \varepsilon_r \psi \text{ (magnetic field representation)},$$

$$\psi = H_x. \qquad (5.247)$$

Using the slowly varying envelope approximation—in this case, that the wave function $\psi(x, y, z)$ propagating in the z direction can be separated into a slowly varying envelope function $\phi(x, y, z)$ and a very fast oscillating phase term $\exp(-j\beta z)$—we get

$$\psi(x, y, z) = \phi(x, y, z) \exp(-j\beta z), \qquad (5.248)$$

where k_0, n_{eff}, and $\beta \ (= n_{\text{eff}} k_0)$ are respectively the wave number in a vacuum, the reference index, and the propagation constant. Assuming the Fresnel approximation

$$\frac{\partial^2 \phi}{\partial z^2} = 0, \qquad (5.249)$$

we reduce Eq. (5.243) to the Fresnel wave equation

$$2j\beta \frac{\partial \phi}{\partial z} = P\phi. \qquad (5.250)$$

It should be noted that $P\psi$ in Eq. (5.250) differs by $-\beta^2$ from that in Eqs. (5.243)–(5.247) and for the quasi-TE mode is expressed as

$$P\phi = \frac{\partial}{\partial x}\left(\frac{1}{\varepsilon_r}\frac{\partial}{\partial x}(\varepsilon_r \phi)\right) + \frac{\partial^2 \phi}{\partial y^2} + (k_0^2 \varepsilon_r - \beta^2)\phi \qquad (5.251)$$

in the electric field representation and as

$$P\phi = \varepsilon_r \frac{\partial}{\partial x}\left(\frac{1}{\varepsilon_r}\frac{\partial \phi}{\partial x}\right) + \frac{\partial^2 \phi}{\partial y^2} + (k_0^2 \varepsilon_r - \beta^2)\phi \qquad (5.252)$$

in the magnetic field representation. For the quasi-TM mode it is expressed as

$$P\phi = \frac{\partial^2 \phi}{\partial x^2} + \frac{\partial}{\partial y}\left(\frac{1}{\varepsilon_r}\frac{\partial}{\partial y}(\varepsilon_r \phi)\right) + (k_0^2 \varepsilon_r - \beta^2)\phi \qquad (5.253)$$

in the electric field representation and as

$$P\phi = \frac{\partial^2 \phi}{\partial x^2} + \varepsilon_r \frac{\partial}{\partial y}\left(\frac{1}{\varepsilon_r}\frac{\partial \phi}{\partial y}\right) + (k_0^2 \varepsilon_r - \beta^2)\phi \qquad (5.254)$$

in the magnetic field representation.

The nonequidistant discretization mesh shown in Fig. 4.3 for the 2D cross-sectional FDM is also used in the lateral directions for the 3D FD-BPM. Here, the subscripts for the lateral positions x and y are respectively p and q and the superscript for position z in the propagation direction is l. Thus, using Eqs. (4.51)–(4.59), we can write the fields, the discretization widths, and the relative permittivity as follows:

$$\phi_{p,q}^l = \phi(x_p, y_q, z_l), \qquad \phi_{p\pm1,q}^l = \phi(x_{p\pm1}, y_q, z_l),$$

$$\phi_{p,q\pm1}^l = \phi(x_p, y_{q\pm1}, z_l), \qquad n = y_q - y_{q-1}, \qquad s = y_{q+1} - y_q,$$

$$e = x_{p+1} - x_p, \qquad w = x_p - x_{p-1}, \qquad \varepsilon_r^l(p,q) = \varepsilon_r(x_p, y_q, z_l). \qquad (5.255)$$

First, we discuss the discretization with respect to x and y. Discretizing Eqs. (5.251)–(5.253), by deduction from Eqs. (4.99), (4.123), (4.133), and (4.143), we get

$$P = \alpha_w^l \phi_{p-1,q}^l + \alpha_e^l \phi_{p+1,q}^l + \alpha_n^l \phi_{p,q-1}^l + \alpha_s^l \phi_{p,q+1}^l$$
$$+ (\alpha_x^l + \alpha_y^l)\phi_{p,q}^l + k_0^2(\varepsilon_r^l(p,q) - n_{\text{eff}}^2)\phi_{p,q}^l,$$

where $\beta = k_0 n_{\text{eff}}$.

Thus, discretizing only the right-hand side of Eq. (5.250) with respect to x and y, we can reduce Eq. (5.250) to

$$2j\beta \frac{\partial \phi}{\partial z} = \alpha_w \phi_{p-1,q} + \alpha_e \phi_{p+1,q} + \alpha_n \phi_{p,q-1} + \alpha_s \phi_{p,q+1}$$
$$+ (\alpha_x + \alpha_y)\phi_{p,q} + k_0^2[\varepsilon_r(p,q) - n_{\text{eff}}^2]\phi_{p,q}. \qquad (5.256)$$

This equation can be rewritten as

$$2j\beta \frac{\partial \phi}{\partial z} = (\alpha_w \phi_{p-1,q} + \alpha_e \phi_{p+1,q} + \alpha_x \phi_{p,q}) \quad \text{(derivative with respect to } x\text{)}$$
$$+ (\alpha_n \phi_{p,q-1} + \alpha_s \phi_{p,q+1} + \alpha_y \phi_{p,q}) \quad \text{(derivative with respect to } y\text{)}$$
$$+ k_0^2 [\varepsilon_r(p,q) - n_{\text{eff}}^2] \phi_{p,q}. \tag{5.257}$$

Now, we move to the discretization of the derivative with respect to z on the left-hand side of Eq. (5.257). A sensitive problem in discretization is the difference centers of the right-hand side and the left-hand side of the equation in the z direction. In the ADI-BPM, the calculation for the step $z \to z + \Delta z$ is divided into two steps, $z \to z + \Delta z/2$ and $z + \Delta z/2 \to z + \Delta z$. In the following, we describe the explicit calculation procedure.

5.4.1 First Step: $z \to z + \Delta z/2 (l \to l + \frac{1}{2})$

The derivative with respect to x on the right-hand side of Eq. (5.257) is written by the implicit FD expression using the unknown fields at $l + \frac{1}{2}$ as

$$\alpha_w^{l+1/2} \phi_{p-1,q}^{l+1/2} + \alpha_e^{l+1/2} \phi_{p+1,q}^{l+1/2} + \alpha_x^{l+1/2} \phi_{p,q}^{l+1/2}. \tag{5.258}$$

The derivative with respect to y, on the other hand, is written by the explicit FD expression with the known fields at l as

$$\alpha_n^l \phi_{p,q-1}^l + \alpha_s^l \phi_{p,q+1}^l + \alpha_y^l \phi_{p,q}^l. \tag{5.259}$$

The remaining part on the right-hand side is expressed by using an average of l and $l + \frac{1}{2}$ as

$$k_0^2 (\varepsilon_r^{l+1/2}(p,q) - n_{\text{eff}}^2) \frac{\phi_{p,q}^l + \phi_{p,q}^{l+1/2}}{2}, \tag{5.260}$$

where $\varepsilon_r^{l+1/2} = (\varepsilon_r^l + \varepsilon_r^{l+1})/2$. With respect to the left-hand side of Eq. (5.257), we get

$$2j\beta \frac{\partial \phi}{\partial z} \to 2j\beta \frac{\phi_{p,q}^{l+1/2} - \phi_{p,q}^l}{\Delta z/2}. \tag{5.261}$$

Thus, from expressions (5.258)–(5.261), we get

$$2j\beta \frac{\phi_{p,q}^{l+1/2} - \phi_{p,q}^{l}}{\Delta z/2} = (\alpha_w^{l+1/2}\phi_{p-1,q}^{l+1/2} + \alpha_e^{l+1/2}\phi_{p+1,q}^{l+1/2} + \alpha_x^{l+1/2}\phi_{p,q}^{l+1/2})$$

$$+ (\alpha_n^l \phi_{p,q-1}^l + \alpha_s^l \phi_{p,q+1}^l + \alpha_y^l \phi_{p,q}^l)$$

$$+ k_0^2(\varepsilon_r^{l+1/2}(p,q) - n_{\text{eff}}^2)\frac{\phi_{p,q}^l + \phi_{p,q}^{l+1/2}}{2}. \quad (5.262)$$

So that the terms on the left- and right-hand sides respectively contain $l + \frac{1}{2}$ and l, we rewrite Eq. (5.262) as

$$-\alpha_w^{l+1/2}\phi_{p-1,q}^{l+1/2} + \left(-\alpha_x^{l+1/2} + \frac{4j\beta}{\Delta z} - \frac{k_0^2}{2}(\varepsilon_r^{l+1/2}(p,q) - n_{\text{eff}}^2)\right)\phi_{p,q}^{l+1/2}$$

$$-\alpha_e^{l+1/2}\phi_{p+1,q}^{l+1/2}$$

$$= \alpha_n^l \phi_{p,q-1}^l + \left(\alpha_y^l + \frac{4j\beta}{\Delta z} + \frac{k_0^2}{2}(\varepsilon_r^{l+1/2}(p,q) - n_{\text{eff}}^2)\right)\phi_{p,q}^l + \alpha_s^l \phi_{p,q+1}^l.$$

(5.263)

5.4.2 Second Step: $z + \Delta z/2 \to z + \Delta z (l + \frac{1}{2} \to l + 1)$

The derivative with respect to y on the right-hand side of Eq. (5.257) is written by the implicit FD expression using the unknown fields at $l + 1$ as

$$\alpha_n^{l+1}\phi_{p,q-1}^{l+1} + \alpha_s^{l+1}\phi_{p,q+1}^{l+1} + \alpha_y^{l+1}\phi_{p,q}^{l+1}. \quad (5.264)$$

The derivative with respect to x, on the other hand, is written by the explicit FD expression using the known fields at $l + \frac{1}{2}$ as

$$\alpha_w^{l+1/2}\phi_{p-1,q}^{l+1/2} + \alpha_e^{l+1/2}\phi_{p+1,q}^{l+1/2} + \alpha_x^{l+1/2}\phi_{p,q}^{l+1/2}. \quad (5.265)$$

The remaining part in the right-hand side is expressed by using an average of $l + \frac{1}{2}$ and $l + 1$ as

$$k_0^2[\varepsilon_r^{l+1/2}(p,q) - n_{\text{eff}}^2]\frac{\phi_{p,q}^{l+1/2} + \phi_{p,q}^{l+1}}{2}. \quad (5.266)$$

With respect to the left-hand side of Eq. (5.257), we get

$$2j\beta \frac{\partial \phi}{\partial z} \rightarrow 2j\beta \frac{\phi_{p,q}^{l+1} - \phi_{p,q}^{l+1/2}}{\Delta z/2} \qquad (5.267)$$

Thus, from Eqs. (5.264)–(5.267), we get

$$2j\beta \frac{\phi_{p,q}^{l+1} - \phi_{p,q}^{l+1/2}}{\Delta z/2} = (\alpha_n^{l+1}\phi_{p,q-1}^{l+1} + \alpha_s^{l+1}\phi_{p,q+1}^{l+1} + \alpha_y^{l+1}\phi_{p,q}^{l+1})$$
$$+ (\alpha_w^{l+1/2}\phi_{p-1,q}^{l+1/2} + \alpha_e^{l+1/2}\phi_{p+1,q}^{l+1/2} + \alpha_x^{l+1/2}\phi_{p,q}^{l+1/2})$$
$$+ k_0^2(\varepsilon_r^{l+1/2}(p,q) - n_{\text{eff}}^2)\frac{\phi_{p,q}^{l+1/2} + \phi_{p,q}^{l+1}}{2}. \qquad (5.268)$$

So that the terms on the left- and right-hand sides respectively contain $l+1$ and $l+\frac{1}{2}$, we rewrite Eq. (5.268) as

$$-\alpha_n^{l+1}\phi_{p,q-1}^{l+1} + \left\{-\alpha_y^{l+1} + \frac{4j\beta}{\Delta z} - \frac{k_0^2}{2}(\varepsilon_r^{l+1/2}(p,q) - n_{\text{eff}}^2)\right\}\phi_{p,q}^{l+1}$$
$$-\alpha_s^{l+1}\phi_{p,q+1}^{l+1}$$
$$= \alpha_w^{l+1/2}\phi_{p-1,q}^{l+1/2} + \left\{\alpha_x^{l+1/2} + \frac{4j\beta}{\Delta z} + \frac{k_0^2}{2}(\varepsilon_r^{l+1/2}(p,q) - n_{\text{eff}}^2)\right\}\phi_{p,q}^{l+1/2}$$
$$+ \alpha_e^{l+1/2}\phi_{p+1,q}^{l+1/2}. \qquad (5.269)$$

As discussed above, since the actual calculation in the two steps of the ADI-BPM is the 2D BPM, it is very numerically efficient. The mode mismatch (obtained from the overlap integral) between the propagating field calculated by the semivectorial ADI-BPM and the initial eigenfield calculated by the scalar FEM is shown in Fig. 5.7 as a function of the propagation distance. Since the propagating field and the initial field are obtained by the semivectorial ADI-BPM and the scalar FEM, the mode mismatch shown in Fig. 5.7 corresponds to the difference between the eigenfield shapes of the FD scheme and the FE scheme. As shown in this figure, the difference between the eigenfields obtained by the FDM and the FEM is very small even though the concepts of the two methods are different.

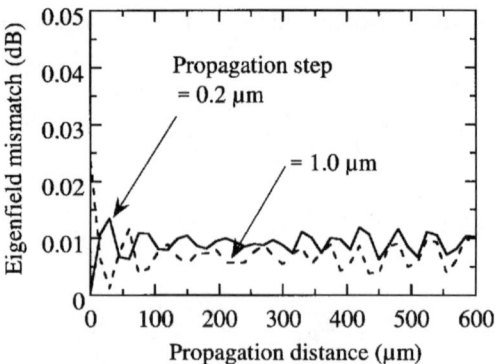

FIGURE 5.7. Eigenfield mismatch between the FEM and FD-BPM.

5.5 THREE-DIMENSIONAL FULLY VECTORIAL ANALYSIS

Since a full discussion of the 3D fully vectorial beam propagation method is beyond the scope of this book, we roughly cover the formulation, emphasizing the relations between the wave equations. For details, refer to Ref. [11].

5.5.1 Wave Equations

As shown in Eqs. (4.17) and (4.18), the 3D vectorial wave equations for the electric field representation are for the x component

$$\frac{\partial}{\partial x}\left(\frac{1}{\varepsilon_r}\frac{\partial}{\partial x}(\varepsilon_r E_x)\right) + \frac{\partial^2 E_x}{\partial y^2} + \frac{\partial^2 E_x}{\partial z^2} + k_0^2 \varepsilon_r E_x + \frac{\partial}{\partial x}\left(\frac{1}{\varepsilon_r}\frac{\partial \varepsilon_r}{\partial y}E_y\right) = 0 \tag{5.270}$$

and for the y component

$$\frac{\partial^2 E_y}{\partial x^2} + \frac{\partial}{\partial y}\left(\frac{1}{\varepsilon_r}\frac{\partial}{\partial y}(\varepsilon_r E_y)\right) + \frac{\partial^2 E_y}{\partial z^2} + k_0^2 \varepsilon_r E_y + \frac{\partial}{\partial y}\left(\frac{1}{\varepsilon_r}\frac{\partial \varepsilon_r}{\partial x}E_x\right) = 0. \tag{5.271}$$

As shown in Eqs. (4.29) and (4.30), the 3D vectorial wave equations for the magnetic field representation are for the x component

$$\frac{\partial^2 H_x}{\partial x^2} + \varepsilon_r \frac{\partial}{\partial y}\left(\frac{1}{\varepsilon_r}\frac{\partial H_x}{\partial y}\right) + \frac{\partial^2 H_x}{\partial z^2} + k_0^2 \varepsilon_r H_x + \frac{1}{\varepsilon_r}\frac{\partial \varepsilon_r}{\partial y}\frac{\partial H_y}{\partial x} = 0 \tag{5.272}$$

and for the *y* component

$$\varepsilon_r \frac{\partial}{\partial x}\left(\frac{1}{\varepsilon_r}\frac{\partial H_y}{\partial x}\right) + \frac{\partial^2 H_y}{\partial y^2} + \frac{\partial^2 H_y}{\partial z^2} + k_0^2 \varepsilon_r H_y + \frac{1}{\varepsilon_r}\frac{\partial \varepsilon_r}{\partial x}\frac{\partial H_x}{\partial y} = 0. \quad (5.273)$$

First, we discuss the electric field representation. Using the SVEA—in this case the wave functions $E_x(x, y, z)$ and $E_y(x, y, z)$ propagating in the z direction can be separated into the slowly varying envelope functions $\tilde{E}_x(x, y, z)$ and $\tilde{E}_y(x, y, z)$ and the vary fast oscillating phase term $\exp(-j\beta z)$—we get

$$E_x(x, y, z) = \tilde{E}_x(x, y, z)\exp(-j\beta z), \quad (5.274)$$

$$E_y(x, y, z) = \tilde{E}_y(x, y, z)\exp(-j\beta z). \quad (5.275)$$

Substituting the second derivatives with respect to z,

$$\frac{\partial^2 E_x}{\partial z^2} = \frac{\partial^2 \tilde{E}_x}{\partial z^2}\exp(-j\beta z) - 2j\beta \frac{\partial \tilde{E}_x}{\partial z}\exp(-j\beta z) - \beta^2 \tilde{E}_x \exp(-j\beta z), \quad (5.276)$$

$$\frac{\partial^2 E_y}{\partial z^2} = \frac{\partial^2 \tilde{E}_y}{\partial z^2}\exp(-j\beta z) - 2j\beta \frac{\partial \tilde{E}_y}{\partial z}\exp(-j\beta z) - \beta^2 \tilde{E}_y \exp(-j\beta z) \quad (5.277)$$

into Eqs. (5.274) and (5.275) and dividing the results by the phase term $\exp(-j\beta z)$, we get the wide-angle equations for the electric field representation:

$$\left(2j\beta - \frac{\partial}{\partial z}\right)\frac{\partial \tilde{E}_x}{\partial z} = P_{xx}\tilde{E}_x + P_{xy}\tilde{E}_y, \quad (5.278)$$

$$\left(2j\beta - \frac{\partial}{\partial z}\right)\frac{\partial \tilde{E}_y}{\partial z} = P_{yy}\tilde{E}_y + P_{yx}\tilde{E}_x. \quad (5.279)$$

The matrix expression for Eqs. (5.278) and (5.279) is

$$\left(2j\beta - \frac{\partial}{\partial z}\right)\frac{\partial}{\partial z}\begin{pmatrix}\tilde{E}_x \\ \tilde{E}_y\end{pmatrix} = \begin{pmatrix}P_{xx} & P_{xy} \\ P_{yx} & P_{yy}\end{pmatrix}\begin{pmatrix}\tilde{E}_x \\ \tilde{E}_y\end{pmatrix}, \quad (5.280)$$

where

$$P_{xx}\tilde{E}_x = \frac{\partial}{\partial x}\left(\frac{1}{\varepsilon_r}\frac{\partial}{\partial x}(\varepsilon_r\tilde{E}_x)\right) + \frac{\partial^2 \tilde{E}_x}{\partial y^2} + (k_0^2\varepsilon_r - \beta^2)\tilde{E}_x, \qquad (5.281)$$

$$P_{xy}\tilde{E}_y = \frac{\partial}{\partial x}\left(\frac{1}{\varepsilon_r}\frac{\partial \varepsilon_r}{\partial y}\tilde{E}_y\right) = \frac{\partial}{\partial x}\left(\frac{1}{\varepsilon_r}\frac{\partial}{\partial y}(\varepsilon_r\tilde{E}_y)\right) - \frac{\partial^2 \tilde{E}_y}{\partial x\, \partial y}, \qquad (5.282)$$

$$P_{yy}\tilde{E}_y = \frac{\partial^2 \tilde{E}_y}{\partial x^2} + \frac{\partial}{\partial y}\left(\frac{1}{\varepsilon_r}\frac{\partial}{\partial y}(\varepsilon_r\tilde{E}_y)\right) + (k_0^2\varepsilon_r - \beta^2)\tilde{E}_y, \qquad (5.283)$$

$$P_{yx}\tilde{E}_x = \frac{\partial}{\partial y}\left(\frac{1}{\varepsilon_r}\frac{\partial \varepsilon_r}{\partial x}\tilde{E}_x\right) = \frac{\partial}{\partial y}\left(\frac{1}{\varepsilon_r}\frac{\partial}{\partial x}(\varepsilon_r\tilde{E}_x)\right) - \frac{\partial^2 \tilde{E}_x}{\partial y\, \partial x}. \qquad (5.284)$$

Using the SVEAs

$$H_x(x,y,z) = \tilde{H}_x(x,y,z)\exp(-j\beta z), \qquad (5.285)$$
$$H_y(x,y,z) = \tilde{H}_y(x,y,z)\exp(-j\beta z), \qquad (5.286)$$

we get the wide-angle wave equation for the magnetic field representation:

$$\left(2j\beta - \frac{\partial}{\partial z}\right)\frac{\partial}{\partial z}\begin{pmatrix}\tilde{H}_x \\ \tilde{H}_y\end{pmatrix} = \begin{pmatrix}P_{xx} & P_{xy} \\ P_{yx} & P_{yy}\end{pmatrix}\begin{pmatrix}\tilde{H}_x \\ \tilde{H}_y\end{pmatrix}, \qquad (5.287)$$

where

$$P_{xx}\tilde{H}_x = \frac{\partial^2 \tilde{H}_x}{\partial x^2} + \varepsilon_r\frac{\partial}{\partial y}\left(\frac{1}{\varepsilon_r}\frac{\partial \tilde{H}_x}{\partial y}\right) + (k_0^2\varepsilon_r - \beta^2)\tilde{H}_x, \qquad (5.288)$$

$$P_{xy}\tilde{H}_y = \frac{1}{\varepsilon_r}\frac{\partial \varepsilon_r}{\partial y}\frac{\partial \tilde{H}_y}{\partial x} = \frac{\partial^2 \tilde{H}_y}{\partial y\, \partial x} - \varepsilon_r\frac{\partial}{\partial y}\left(\frac{1}{\varepsilon_r}\frac{\partial \tilde{H}_y}{\partial x}\right), \qquad (5.289)$$

$$P_{yy}\tilde{H}_y = \varepsilon_r\frac{\partial}{\partial x}\left(\frac{1}{\varepsilon_r}\frac{\partial \tilde{H}_y}{\partial x}\right) + \frac{\partial^2 \tilde{H}_y}{\partial y^2} + (k_0^2\varepsilon_r - \beta^2)\tilde{H}_y, \qquad (5.290)$$

$$P_{yx}\tilde{H}_x = \frac{1}{\varepsilon_r}\frac{\partial \varepsilon_r}{\partial x}\frac{\partial \tilde{H}_x}{\partial y} = \frac{\partial^2 \tilde{H}_x}{\partial x\, \partial y} - \varepsilon_r\frac{\partial}{\partial x}\left(\frac{1}{\varepsilon_r}\frac{\partial \tilde{H}_x}{\partial y}\right). \qquad (5.291)$$

Neglecting the interaction terms P_{xy} and P_{yx} in Eqs. (5.280) and (5.287), we get the semivectorial wave equations. Neglecting the second terms on the left-hand sides of Eqs. (5.280) and (5.287), we get the vectorial wave equations based on the Fresnel approximation.

5.5.2 Finite-Difference Expressions

Here, we discuss the fully vectorial FD-BPM using as an example the vectorial wave equations (5.287)–(5.291) for the magnetic field representation. For simplicity, the equidistant discretizations Δx and Δy are assumed in the x and y directions.

Since P_{xx} is an operator corresponding to the semivectorial analysis for $\tilde{H}_x$,

$$P_{xx}\tilde{H}_x = \alpha_w^{xx}\tilde{H}_x(p-1,q) + \alpha_e^{xx}\tilde{H}_x(p+1,q)$$
$$+ \alpha_n^{xx}\tilde{H}_x(p,q-1) + \alpha_s^{xx}\tilde{H}_x(p,q+1)$$
$$+ (\alpha_x^{xx} + \alpha_y^{xx})\tilde{H}_x(p,q) + \{k_0^2\varepsilon_r(p,q) - \beta^2\}\tilde{H}_x(p,q), \quad (5.292)$$

where, from Eqs. (4.143)–(4.149), we get

$$\alpha_w^{xx} = \frac{1}{(\Delta x)^2}, \quad (5.293)$$

$$\alpha_e^{xx} = \frac{1}{(\Delta x)^2}, \quad (5.294)$$

$$\alpha_n^{xx} = \frac{1}{(\Delta y)^2} \frac{2\varepsilon_r(p,q)}{\varepsilon_r(p,q) + \varepsilon_r(p,q-1)}, \quad (5.295)$$

$$\alpha_s^{xx} = \frac{1}{(\Delta y)^2} \frac{2\varepsilon_r(p,q)}{\varepsilon_r(p,q) + \varepsilon_r(p,q+1)}, \quad (5.296)$$

$$\alpha_x^{xx} = -\frac{2}{(\Delta x)^2} = -\alpha_e^{xx} - \alpha_w^{xx}, \quad (5.297)$$

$$\alpha_y^{xx} = -\frac{1}{(\Delta y)^2} \frac{2\varepsilon_r(p,q)}{\varepsilon_r(p,q) + \varepsilon_r(p,q-1)} - \frac{1}{(\Delta y)^2} \frac{2\varepsilon_r(p,q)}{\varepsilon_r(p,q) + \varepsilon_r(p,q+1)}$$
$$= -\alpha_n^{xx} - \alpha_s^{xx}. \quad (5.298)$$

On the other hand,

$$\begin{aligned}P_{yy}\tilde{H}_y = {} & \alpha_w^{yy}\tilde{H}_y(p-1,q) + \alpha_e^{yy}\tilde{H}_y(p+1,q) \\ & + \alpha_n^{yy}\tilde{H}_y(p,q-1) + \alpha_s^{yy}\tilde{H}_y(p,q+1) \\ & + (\alpha_x^{yy} + \alpha_y^{yy})\tilde{H}_y(p,q) + \{k_0^2\varepsilon_r(p,q) - \beta^2\}\tilde{H}_y(p,q), \end{aligned} \quad (5.299)$$

where, from Eqs (4.123) to (4.129), we get

$$\alpha_w^{yy} = \frac{1}{(\Delta x)^2} \frac{2\varepsilon_r(p,q)}{\varepsilon_r(p,q) + \varepsilon_r(p-1,q)}, \quad (5.300)$$

$$\alpha_e^{yy} = \frac{1}{(\Delta x)^2} \frac{2\varepsilon_r(p,q)}{\varepsilon_r(p,q) + \varepsilon_r(p+1,q)}, \quad (5.301)$$

$$\alpha_n^{yy} = \frac{1}{(\Delta y)^2}, \quad (5.302)$$

$$\alpha_s^{yy} = \frac{1}{(\Delta y)^2}, \quad (5.303)$$

$$\begin{aligned}\alpha_x^{yy} = {} & -\frac{1}{(\Delta x)^2} \frac{2\varepsilon_r(p,q)}{\varepsilon_r(p,q) + \varepsilon_r(p-1,q)} \\ & -\frac{1}{(\Delta x)^2} \frac{2\varepsilon_r(p,q)}{\varepsilon_r(p,q) + \varepsilon_r(p+1,q)} = -\alpha_w^{xx} - \alpha_e^{xx}, \end{aligned} \quad (5.304)$$

$$\alpha_y^{yy} = -\frac{2}{(\Delta y)^2} = -\alpha_n^{yy} - \alpha_s^{yy}. \quad (5.305)$$

Although the derivation is not shown here, the FD expressions for the interaction terms $P_{xy}\tilde{H}_y$ and $P_{yx}\tilde{H}_x$ (for the interaction between

$\tilde{H}_x$ and $\tilde{H}_y$) are

$$P_{xy}\tilde{H}_y = \frac{1}{4\,\Delta x\,\Delta y}\left(1 - \frac{\varepsilon_r(p,q)}{\varepsilon_r(p,q+1)}\right)\tilde{H}_y(p+1,q+1)$$
$$- \left(1 - \frac{\varepsilon_r(p,q)}{\varepsilon_r(p,q-1)}\right)\tilde{H}_y(p+1,q-1)$$
$$- \left(1 - \frac{\varepsilon_r(p,q)}{\varepsilon_r(p,q+1)}\right)\tilde{H}_y(p-1,q+1)$$
$$+ \left(1 - \frac{\varepsilon_r(p,q)}{\varepsilon_r(p,q-1)}\right)\tilde{H}_y(p-1,q-1), \qquad (5.306)$$

$$P_{yx}\tilde{H}_x = \frac{1}{4\,\Delta x\,\Delta y}\left(1 - \frac{\varepsilon_r(p,q)}{\varepsilon_r(p+1,q)}\right)\tilde{H}_x(p+1,q+1)$$
$$- \left(1 - \frac{\varepsilon_r(p,q)}{\varepsilon_r(p-1,q)}\right)\tilde{H}_x(p-1,q+1)$$
$$- \left(1 - \frac{\varepsilon_r(p,q)}{\varepsilon_r(p+1,q)}\right)\tilde{H}_x(p+1,q-1)$$
$$+ \left(1 - \frac{\varepsilon_r(p,q)}{\varepsilon_r(p-1,q)}\right)\tilde{H}_x(p-1,q-1). \qquad (5.307)$$

The final FD beam propagation equations to be solved can be obtained by applying the FD procedure to the derivative with respect to the z coordinate.

PROBLEMS

1. Using a plane wave, evaluate the difference error for the FD approximation used in the FD-BPM.

ANSWER

When we use the center difference scheme, the second derivative of the wave function $\phi(x)$ with respect to x is

$$\frac{d^2\phi_i}{dx^2} = \frac{\phi_{i+1} - 2\phi_i + \phi_{i-1}}{(\Delta x)^2}, \qquad (P5.1)$$

where Δx is the discretization width and i is the number of the node at the difference center.

If we assume that wave function $\phi(x)$ is a plane wave

$$\phi(x) = \phi_0 \exp(-jk_x x) = \phi_0 \exp(-jk_x \Delta x\, i), \tag{P5.2}$$

the left- and right-hand sides of Eq. (5.1) can be respectively rewritten as

$$-k_x^2 \phi_i \tag{P5.3}$$

and

$$\frac{1}{(\Delta x)^2} \phi_i [\exp(-jk_x \Delta x) - 2 + \exp(-jk_x \Delta x)]$$

$$= \frac{1}{(\Delta x)^2} [2 \cos(k_x \Delta x) - 2] \phi_i$$

$$= \frac{2}{(\Delta x)^2} (\cos(k_x \Delta x) - 1) \phi_i$$

$$= \frac{2}{(\Delta x)^2} \left[-2 \sin^2 \left(\frac{k_x \Delta x}{2} \right) \right] \phi_i.$$

Using these two factors, we get the relative error ε:

$$\varepsilon = \frac{\{k_x^2 - [4/(\Delta x)^2] \sin^2(k_x \Delta x/2)\} \phi_i}{k_x^2 \phi_i} = \frac{1}{k_x^2} \left\{ k_x^2 - \left[\frac{2}{\Delta x} \sin\left(\frac{k_x \Delta x}{2} \right) \right]^2 \right\}. \tag{P5.4}$$

When the discretization width Δx is very small, $\sin(k_x \Delta x/2)$ can be approximated as $k_x \Delta x/2$. Thus, the relative error ε becomes zero as $\Delta x \to 0$.

2. Figure P5.1 shows a simple example of an analysis region for a 1D FD-BPM. Nodes 1–4 are inside the analysis window, and nodes 0–5 are outside the window. Show the form of the matrix equation for this example.

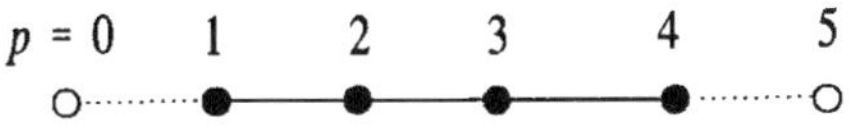

FIGURE P5.1. Simple example of a 1D FD-BPM.

ANSWER

$$\begin{pmatrix} B'(1) & C(1) & & \\ A(2) & B(2) & C(2) & \\ & A(3) & B(3) & C(3) \\ & & A(4) & B'(4) \end{pmatrix} \begin{pmatrix} \phi_1^{l+1} \\ \phi_2^{l+1} \\ \phi_3^{l+1} \\ \phi_4^{l+1} \end{pmatrix} = \begin{pmatrix} D(1) \\ D(2) \\ D(3) \\ D(4) \end{pmatrix}, \quad (P5.5)$$

where $B'(1)$ and $B'(4)$ include the boundary conditions due to nodes 0 and 5. For explicit forms of the coefficients on the left-hand side and of $D(1)$ to $D(4)$ on the right-hand side, see Eqs. (5.178)–(5.181) and Eqs. (5.186) and (5.193).

3. The 3D semivectorial analysis shown in Section 5.3 was based on the ADI method. Discuss the procedure for a 3D analysis not using the ADI method.

ANSWER

The starting equation is Eq. (5.256). Using a procedure similar to the one specified in Eqs. (5.131)–(5.133), we get

$$2j\beta \frac{\phi_{p,q}^{l+1} - \phi_{p,q}^l}{\Delta z}$$

$$= \frac{1}{2}(\alpha_w^{l+1}\phi_{p-1,q}^{l+1} + \{\alpha_x^{l+1} + k_0^2[\varepsilon_r^{l+1}(p,q) - n_{\text{eff}}^2]\}\phi_{p,q}^{l+1}$$

$$+ \alpha_e^{l+1}\phi_{p+1,q}^{l+1} + \alpha_n^{l+1}\phi_{p,q-1}^{l+1} + \alpha_s^{l+1}\phi_{p,q+1}^{l+1} + \alpha_y^{l+1}\phi_{p,q}^{l+1})$$

$$+ \frac{1}{2}(\alpha_w^l\phi_{p-1,q}^l + \{\alpha_x^l + k_0^2[\varepsilon_r^l(p,q) - n_{\text{eff}}^2]\}\phi_{p,q}^l + \alpha_e^l\phi_{p+1,q}^l$$

$$+ \alpha_n^l\phi_{p,q-1}^l + \alpha_s^l\phi_{p,q+1}^l + \alpha_y^l\phi_{p,q}^l),$$

where the superscripts $l+1$ and l respectively correspond to the unknown and the known quantities.

Thus, we get the 3D FD-BPM expression

$$-\alpha_w^{l+1}\phi_{p-1,q}^{l+1} + \left\{-\alpha_x^{l+1} - \alpha_y^{l+1} + \frac{4j\beta}{\Delta z} - k_0^2[\varepsilon_r^{l+1}(p,q) - n_{\text{eff}}^2]\right\}\phi_{p,q}^{l+1}$$

$$-\alpha_e^{l+1}\phi_{p+1,q}^{l+1} - \alpha_n^{l+1}\phi_{p,q-1}^{l+1} - \alpha_s^{l+1}\phi_{p,q+1}^{l+1}$$

$$= \alpha_w^l \phi_{p-1,q}^l + \left\{\alpha_x^l + \alpha_y^l + \frac{4j\beta}{\Delta z} + k_0^2[\varepsilon_r^l(p,q) - n_{\text{eff}}^2]\right\}\phi_{p,q}^l$$

$$+ \alpha_e^l \phi_{p+1,q}^l + \alpha_n^l \phi_{p,q-1}^l + \alpha_s^l \phi_{p,q+1}^l. \tag{P5.6}$$

REFERENCES

[1] M. D. Feit and J. A. Freck, Jr., "Light propagation in graded-index optical fibers," *Appl. Opt.*, vol. 17, pp. 3990–3998, 1978.

[2] L. Thylén, "The beam propagation method: An analysis of its applicability," *Opt. Quantum Electron*, vol. 15, pp. 433–439, 1983.

[3] J. Yamauchi, J. Shibayama, and H. Nakano, "Beam propagation method using Padé approximant operators," *Trans. IEICE Jpn.*, vol. J77-C-I, pp. 490–494, 1994.

[4] Y. Chung and N. Dagli, "Assessment of finite difference beam propagation," *IEEE J. Quantum Electron.*, vol. 26, pp. 1335–1339, 1990.

[5] G. R. Hadley, "Wide-angle beam propagation using Padé approximant operators," *Opt. Lett.*, vol. 17, pp. 1426–1428, 1992.

[6] G. R. Hadley, "A multistep method for wide angle beam propagation," *Integrated Photon. Res.*, vol. ITu I5-1, pp. 387–391, 1993.

[7] J. Yamauchi, J. Shibayama, and H. Nakano, "Modified finite-difference beam propagation method on the generalized Douglas scheme for variable coefficients," *IEEE Photon. Technol. Lett.*, vol. 7, pp. 661–663, 1995.

[8] J. Yamauchi, T. Ando, and H. Nakano, "Beam-propagation analysis of optical fibres by alternating direction implicit method," *Electron. Lett.*, vol. 27, pp. 1663–1665, 1991.

[9] J. Yamauchi, T. Ando, and H. Nakano, "Propagating beam analysis by alternating-direction implicit finite-difference method," *Trans. IEICE Jpn.*, vol. J75-C-I, pp. 148–154, 1992 (in Japanese).

[10] P. L. Liu and B. J. Li, "Study of form birefringence in waveguide devices using the semivectorial beam propagation method," *IEEE Photon. Technol. Lett.*, vol. 3, pp. 913–915, 1991.

[11] W. P. Huang and C. L. Xu, "Simulation of three-dimensional optical waveguides by a full-vector beam propagation method," *IEEE J. Quantum Electron.*, vol. 29, pp. 2639–2649, 1993.

[12] M. Koshiba and Y. Tsuji, "A wide-angle finite element beam propagation method," *IEEE Photon. Technol. Lett.*, vol. 8, pp. 1208–1210, 1996.

[13] G. R. Hadley, "Transparent boundary condition for beam propagation," *Opt. Lett.*, vol. 16, pp. 624–626, 1992.

[14] W. H. Press, S. A. Teukolsky, W. T. Vetterling, and B. P. Flannery, *Numerical Recipes*, Cambridge University Press, New York, 1992.

[15] K. Kawano, M. Kohtoku, M. Wada, H. Okamoto, Y. Itaya, and M. Naganuma, "Design of spotsize-converter-integrated laser diode (SS-LD) with a lateral taper, thin-film core and ridge in the 1.3 µm wavelength region based on the 3-D BPM," *IEEE J. Select. Top. Quantum Electron.*, vol. 2, pp. 348–354, 1996.

[16] K. Kawano, M. Kohtoku, H. Okamoto, Y. Itaya, and M. Naganuma, "Coupling and conversion characteristics of spot-size converter integrated laser diodes," *IEEE J. Select. Top. Quantum Electron.*, vol. 3, pp. 1351–1360, 1997.

[17] K. Kawano, M. Kohtoku, N. Yoshimoto, S. Sekine, and Y. Noguchi, "2 × 2 InGaAlAs/InAlAs multiple quantum well (MQW) directional coupler waveguide switch modules integrated with spot-size converters," *Electron. Lett.*, vol. 30, pp. 353–354, 1994.

CHAPTER 6

FINITE-DIFFERENCE TIME-DOMAIN METHOD

In the preceding chapters, steady-state wave equations were solved in which the derivative with respect to the time t (i.e., $\partial/\partial t$) was replaced by $j\omega$. In this chapter, we discuss the finite-difference time-domain method (FD-TDM), which was developed by Yee [1] and which directly solves time-dependent Maxwell equations. The FD-TDM was originally proposed for electromagnetic waves with long wavelengths, such as microwaves, because the spatial discretization it requires is small ($\frac{1}{10} - \frac{1}{20}$ of the wavelength). As the FD-TDM is an explicit scheme, the time step in the calculation is defined by the spatial discretization width. Thus, the time step in the optical waveguide analysis is extremely short when wavelengths are of micrometer order. The amount of required memory is enormous for 3D structures, but the method is readily applicable to 2D structures. Finite-difference TDM CAD software suitable for microwave wavelengths as well as optical wavelengths is available on the market.

6.1 DISCRETIZATION OF ELECTROMAGNETIC WAVES

The 3D formulation is shown here because it is more versatile than the 2D formulation, which can be easily obtained from the 3D formulation.

FINITE-DIFFERENCE TIME-DOMAIN METHOD

The time-dependent Maxwell equations are

$$-\mu_0 \frac{\partial \mathbf{H}}{\partial t} = \nabla \times \mathbf{E}, \tag{6.1}$$

$$\varepsilon_0 \varepsilon_r \frac{\partial \mathbf{E}}{\partial t} = \nabla \times \mathbf{H}, \tag{6.2}$$

and using Eqs. (2.5)–(2.10), we can write their component representations as follows:

$$-\mu_0 \frac{\partial H_x}{\partial t} = \frac{\partial E_z}{\partial y} - \frac{\partial E_y}{\partial z}, \tag{6.3}$$

$$-\mu_0 \frac{\partial H_y}{\partial t} = \frac{\partial E_x}{\partial z} - \frac{\partial E_z}{\partial x}, \tag{6.4}$$

$$-\mu_0 \frac{\partial H_z}{\partial t} = \frac{\partial E_y}{\partial x} - \frac{\partial E_x}{\partial y}, \tag{6.5}$$

$$\varepsilon_0 \varepsilon_r \frac{\partial E_x}{\partial t} = \frac{\partial H_z}{\partial y} - \frac{\partial H_y}{\partial z}, \tag{6.6}$$

$$\varepsilon_0 \varepsilon_r \frac{\partial E_y}{\partial t} = \frac{\partial H_x}{\partial z} - \frac{\partial H_z}{\partial x}, \tag{6.7}$$

$$\varepsilon_0 \varepsilon_r \frac{\partial E_z}{\partial t} = \frac{\partial H_y}{\partial x} - \frac{\partial H_x}{\partial y}. \tag{6.8}$$

When we assume that Δx, Δy, and Δz are spatial discretizations and that Δt is a time step, the function $F(x, y, z, t)$ is discretized as

$$F^n(i, j, k) = F(i\Delta x, j\Delta y, k\Delta z, n\Delta t) = F(x, y, z, t). \tag{6.9}$$

Figure 6.1 shows what we call the Yee lattice [1]. Using α to represent a spatial coordinate such as x, y, and z, we define

$$E_\alpha = \begin{cases} \text{spatial coordinate } \alpha\text{: half-integer,} \\ \text{the other spatial ones: integer,} \\ \text{time: integer,} \end{cases} \tag{6.10}$$

$$H_\alpha = \begin{cases} \text{spatial coordinate } \alpha\text{: integer,} \\ \text{the other spatial ones: half-integer,} \\ \text{time: half-integer,} \end{cases} \tag{6.11}$$

Difference centers play important roles when Eqs. (6.3)–(6.8) are discretized, and here we investigate Eq. (6.3) as an example. Since the spatial difference centers of the left-hand side of the equation are the same

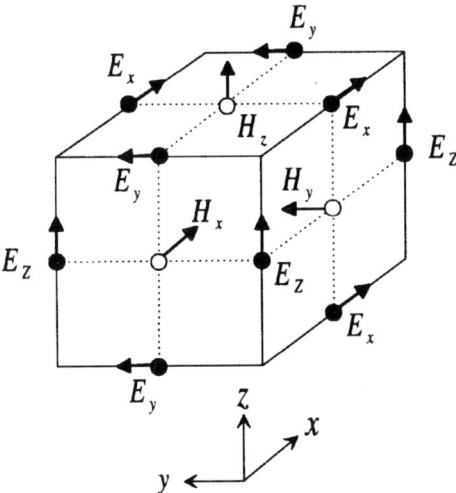

FIGURE 6.1. Yee lattice.

as those for H_x, the spatial difference centers in the x, y, and z directions are respectively found to be $x = i\,\Delta x$, $y = (j + \tfrac{1}{2})\,\Delta y$, and $z = (k + \tfrac{1}{2})\,\Delta z$. Since the time difference center of the right-hand side of the equation is the same as that for the electric fields E_y and E_z, we can write $t = n\,\Delta t$. Here, i, j, k, and n are integers. In a similar manner, we can also obtain the following difference centers of the spatial coordinates and the time for Eqs. (6.3)–(6.8).

Eq. (6.3): $\quad x = i\,\Delta x, \qquad y = (j + \tfrac{1}{2})\,\Delta y, \quad z = (k + \tfrac{1}{2})\,\Delta z,$
$\qquad\qquad t = n\,\Delta t,$ (6.12)

Eq. (6.4): $\quad x = (i + \tfrac{1}{2})\,\Delta x, \quad y = j\,\Delta y, \qquad z = (k + \tfrac{1}{2})\,\Delta z,$
$\qquad\qquad t = n\,\Delta t,$ (6.13)

Eq. (6.5): $\quad x = (i + \tfrac{1}{2})\,\Delta x, \quad y = (j + \tfrac{1}{2})\,\Delta y, \quad z = k\,\Delta z,$
$\qquad\qquad t = n\,\Delta t,$ (6.14)

Eq. (6.6): $\quad x = (i + \tfrac{1}{2})\,\Delta x, \quad y = j\,\Delta y, \qquad z = k\,\Delta z,$
$\qquad\qquad t = (n + \tfrac{1}{2})\,\Delta t,$ (6.15)

Eq. (6.7): $\quad x = i\,\Delta x, \qquad y = (j + \tfrac{1}{2})\,\Delta y, \quad z = k\,\Delta z,$
$\qquad\qquad t = (n + \tfrac{1}{2})\,\Delta t,$ (6.16)

Eq. (6.8): $\quad x = i\,\Delta x, \qquad y = j\,\Delta y, \qquad z = (k + \tfrac{1}{2})\,\Delta z,$
$\qquad\qquad t = (n + \tfrac{1}{2})\,\Delta t.$ (6.17)

The 3D finite-difference time-domain expressions for Eqs. (6.3)–(6.8) can be obtained by discretizing them on the basis of the difference centers (6.12)–(6.17). Again, we investigate Eq. (6.3) as an example. The difference centers x and t are both integers and the difference centers y and z are both half-integers. Thus, for the left-hand side of Eq. (6.3), we get

$$-\frac{\mu_0}{\Delta t}[H_x^{n+1/2}(i,j+\tfrac{1}{2},k+\tfrac{1}{2}) - H_x^{n-1/2}(i,j+\tfrac{1}{2},k+\tfrac{1}{2})]. \tag{6.18}$$

For the right-hand side, we get

$$\frac{1}{\Delta y}[E_z^n(i,j+1,k+\tfrac{1}{2}) - E_z^n(i,j,k+\tfrac{1}{2})]$$
$$-\frac{1}{\Delta z}[E_y^n(i,j+\tfrac{1}{2},k+1) - E_y^n(i,j+\tfrac{1}{2},k)]. \tag{6.19}$$

Using expressions (6.18) and (6.19), we get the following finite-difference time-domain expression for Eq. (6.3):

$$H_x^{n+1/2}(i,j+\tfrac{1}{2},k+\tfrac{1}{2}) = H_x^{n-1/2}(i,j+\tfrac{1}{2},k+\tfrac{1}{2})$$
$$-\frac{\Delta t}{\mu_0}\left\{\frac{1}{\Delta y}[E_z^n(i,j+1,k+\tfrac{1}{2}) - E_z^n(i,j,k+\tfrac{1}{2})]\right.$$
$$\left.-\frac{1}{\Delta z}[E_y^n(i,j+\tfrac{1}{2},k+1) - E_y^n(i,j+\tfrac{1}{2},k)]\right\}. \tag{6.20}$$

Through the same procedure, we get the following finite-difference time-domain expressions for the y and z components of the magnetic fields:

$$H_y^{n+1/2}(i+\tfrac{1}{2},j,k+\tfrac{1}{2}) = H_y^{n-1/2}(i+\tfrac{1}{2},j,k+\tfrac{1}{2})$$
$$-\frac{\Delta t}{\mu_0}\left\{\frac{1}{\Delta z}[E_x^n(i+\tfrac{1}{2},j,k+1) - E_x^n(i+\tfrac{1}{2},j,k)]\right.$$
$$\left.-\frac{1}{\Delta x}[E_z^n(i+1,j,k+\tfrac{1}{2}) - E_z^n(i,j,k+\tfrac{1}{2})]\right\}. \tag{6.21}$$

and

$$H_z^{n+1/2}(i+\tfrac{1}{2},j+\tfrac{1}{2},k) = H_z^{n-1/2}(i+\tfrac{1}{2},j+\tfrac{1}{2},k)$$
$$-\frac{\Delta t}{\mu_0}\left\{\frac{1}{\Delta x}[E_y^n(i+1,j+\tfrac{1}{2},k) - E_y^n(i,j+\tfrac{1}{2},k)]\right.$$
$$\left.-\frac{1}{\Delta y}[E_x^n(i+\tfrac{1}{2},j+1,k) - E_x^n(i+\tfrac{1}{2},j,k)]\right\}. \quad (6.22)$$

Next, we discretize Eq. (6.6). According to expression (6.15), the difference centers of x and t are both half-integers and the difference centers y and z are both integers. Thus, for the left-hand side of Eq. (6.6), we get

$$\frac{\varepsilon_0 \varepsilon_r}{\Delta t}[E_x^{n+1}(i+\tfrac{1}{2},j,k) - E_x^n(i+\tfrac{1}{2},j,k)]. \quad (6.23)$$

For the right-hand side, we get

$$\frac{1}{\Delta y}[H_z^{n+1/2}(i+\tfrac{1}{2},j+\tfrac{1}{2},k) - H_z^{n+1/2}(i+\tfrac{1}{2},j-\tfrac{1}{2},k)]$$
$$-\frac{1}{\Delta z}[H_y^{n+1/2}(i+\tfrac{1}{2},j,k+\tfrac{1}{2}) - H_y^{n+1/2}(i+\tfrac{1}{2},j,k-\tfrac{1}{2})]. \quad (6.24)$$

Using expressions (6.23) and (6.24), we get the following finite-difference time-domain expression for Eq. (6.6):

$$E_x^{n+1}(i+\tfrac{1}{2},j,k) = E_x^n(i+\tfrac{1}{2},j,k)$$
$$+\frac{\Delta t}{\varepsilon_0 \varepsilon_r}\left\{\frac{1}{\Delta y}[H_z^{n+1/2}(i+\tfrac{1}{2},j+\tfrac{1}{2},k)\right.$$
$$- H_z^{n+1/2}(i+\tfrac{1}{2},j-\tfrac{1}{2},k)]$$
$$-\frac{1}{\Delta z}[H_y^{n+1/2}(i+\tfrac{1}{2},j,k+\tfrac{1}{2})$$
$$\left.- H_y^{n+1/2}(i+\tfrac{1}{2},j,k-\tfrac{1}{2})]\right\}. \quad (6.25)$$

Through the same procedure, we get the following finite-difference time-domain expressions for the y and z components of the electric fields:

$$E_y^{n+1}(i,j+\tfrac{1}{2},k) = E_y^n(i,j+\tfrac{1}{2},k)$$
$$+ \frac{\Delta t}{\varepsilon_0 \varepsilon_r} \left\{ \frac{1}{\Delta z}[H_x^{n+1/2}(i,j+\tfrac{1}{2},k+\tfrac{1}{2}) \right.$$
$$- H_x^{n+1/2}(i,j+\tfrac{1}{2},k-\tfrac{1}{2})]$$
$$- \frac{1}{\Delta x}[H_z^{n+1/2}(i+\tfrac{1}{2},j+\tfrac{1}{2},k)$$
$$\left. - H_z^{n+1/2}(i-\tfrac{1}{2},j+\tfrac{1}{2},k)] \right\} \quad (6.26)$$

and

$$E_z^{n+1}(i,j,k+\tfrac{1}{2}) = E_z^n(i,j,k+\tfrac{1}{2})$$
$$+ \frac{\Delta t}{\varepsilon_0 \varepsilon_r} \left\{ \frac{1}{\Delta x}[H_y^{n+1/2}(i+\tfrac{1}{2},j,k+\tfrac{1}{2}) \right.$$
$$- H_y^{n+1/2}(i-\tfrac{1}{2},j,k+\tfrac{1}{2})]$$
$$- \frac{1}{\Delta y}[H_x^{n+1/2}(i,j+\tfrac{1}{2},k+\tfrac{1}{2})$$
$$\left. - H_x^{n+1/2}(i,j-\tfrac{1}{2},k+\tfrac{1}{2})] \right\}. \quad (6.27)$$

Magnetic fields $H_\alpha^{n+1/2}$ with the half-integer time step $(n+1/2)\Delta t$ are calculated first from Eqs. (6.20)–(6.22) by using the electric fields with the integer time step $n\,\Delta t$. Then those fields are used to calculate the electric fields E_α^{n+1} with the integer time step $(n+1)\Delta t$ by using Eqs. (6.25)–(6.27). Repeating these two steps, we can calculate the time evolution of the electric and magnetic fields directly.

It should be noted that the relative permittivity at the interface between two media is approximated better by using $(\varepsilon_{r1}+\varepsilon_{r2})/2$ than by using only ε_{r1} or ε_{r2}.

6.2 STABILITY CONDITION

In an explicit scheme such as the FD-TDM, the time step Δt in the calculation is restricted by the spatial discretization. For simplicity in discussing the stability condition here, we will use the 1D scalar Helmholtz equation

$$\frac{\partial^2 \phi}{\partial x^2} - \varepsilon \mu_0 \frac{\partial^2 \phi}{\partial t^2} = 0, \tag{6.28}$$

where ϕ is a 1D wave function that designates the time-dependent field. Using β_x as the x-directed propagation constant, we express this wave function as

$$\begin{aligned}\phi(x, t) &= \exp(j\beta_x x)\exp(\alpha t) \\ &= \exp(j\beta_x p\, \Delta x)\exp(\alpha n\, \Delta t) \\ &= \exp(j\beta_x p\, \Delta x)\xi^n,\end{aligned} \tag{6.29}$$

where $\xi = \exp(\alpha\, \Delta t)$. Thus, if the field is to be stable, ξ has to satisfy the condition

$$|\xi| \leq 1. \tag{6.30}$$

Substituting Eq. (6.29) into (6.28), we get

$$\frac{1}{(\Delta x)^2}\{\exp[j\beta_x(p+1)\,\Delta x]\xi^n - 2\,\exp(j\beta_x p\,\Delta x)\xi^n + \exp[j\beta_x(p-1)\,\Delta x]\xi^n\}$$
$$- \frac{\varepsilon\mu_0}{(\Delta t)^2}\{\exp(j\beta_x p\,\Delta x)\xi^{n+1} - 2\,\exp(j\beta_x p\,\Delta x)\xi^n + \exp(j\beta_x p\,\Delta x)\xi^{n-1}\} = 0.$$

Dividing this equation by $\exp(j\beta_x p\,\Delta x)\xi^n$, we reduce it to

$$\frac{1}{(\Delta x)^2}\{\exp(j\beta_x\,\Delta x) - 2 + \exp(-j\beta_x\,\Delta x)\} - \frac{\varepsilon\mu_0}{(\Delta t)^2}(\xi - 2 + \xi^{-1}) = 0. \tag{6.31}$$

Dividing Eq. (6.31) by $\varepsilon\mu_0/(\Delta t)^2 \xi$ and considering that the first term of Eq. (6.31) can be rewritten as

$$\{\exp(j\beta_x \Delta x) - 2 + \exp(-j\beta_x \Delta x)\} = 2(\cos(\beta_x \Delta x) - 1)$$
$$= -4\sin^2\left(\beta_x \frac{\Delta x}{2}\right),$$

we get

$$(\xi^2 - 2\xi + 1) - \frac{(\Delta t)^2}{\varepsilon\mu_0}\left[-\frac{4}{(\Delta x)^2}\sin^2\left(\beta_x \frac{\Delta x}{2}\right)\right]\xi = 0$$

and therefore

$$\xi^2 - 2A + 1 = 0, \tag{6.32}$$

where the parameter A is defined by

$$A = -\frac{2(\Delta t)^2}{\varepsilon\mu_0}\frac{1}{(\Delta x)^2}\sin^2\left(\beta_x \frac{\Delta x}{2}\right) + 1. \tag{6.33}$$

The roots of Eq. (6.32) are

$$\xi_1 = A + \sqrt{A^2 - 1}, \tag{6.34}$$

$$\xi_2 = A - \sqrt{A^2 - 1}. \tag{6.35}$$

Because $|\xi| \leq 1$ and $0 \leq \sin^2\theta$, we get the relation

$$A = -\frac{2(\Delta t)^2}{\varepsilon\mu_0}\frac{1}{(\Delta x)^2}\sin^2\left(\beta_x \frac{\Delta x}{2}\right) + 1 \leq 1. \tag{6.36}$$

We can thus specify the stability condition in terms of A:

Case $A < -1$. Since, according to Eq. (6.35), $1 < |\xi_2|$, the field is unstable.

Case $-1 \leq A \leq 1$. Since ξ_1 and ξ_2 can be expressed as

$$\xi_1 = A + \sqrt{A^2 - 1} = A + j\sqrt{1 - A^2}, \tag{6.37}$$

$$\xi_2 = A - \sqrt{A^2 - 1} = A - j\sqrt{1 - A^2}, \tag{6.38}$$

their absolute values can be expressed as

$$|\xi_1| = |\xi_2| = A^2 + (1 - A^2) = 1. \tag{6.39}$$

Thus, the field is stable when

$$-1 \leq A \leq 1. \tag{6.40}$$

Relation (6.40) can be interpreted as imposing the following restriction on the time step (see Problem 1):

$$\Delta t \leq \sqrt{\varepsilon \mu_0} \left(\frac{1}{(\Delta x)^2} \right)^{-1/2} = \frac{\sqrt{\varepsilon_r}}{c_0} \left(\frac{1}{(\Delta x)^2} \right)^{-1/2} = \frac{1}{v} \left(\frac{1}{(\Delta x)^2} \right)^{-1/2}, \tag{6.41}$$

where Δx is the spatial discretization width and Δt is the time step and where c_0, ε_r, and $v = c_0/\sqrt{\varepsilon_r}$ are respectively the velocity of the light in a vacuum, the relative permittivity of the medium, and the velocity of the light in the medium. Equation (6.41) is for a 1D structure, and the corresponding restriction for a 3D structure is

$$\Delta t \leq \frac{1}{v} \left(\frac{1}{(\Delta x)^2} + \frac{1}{(\Delta y)^2} + \frac{1}{(\Delta z)^2} \right)^{-1/2}. \tag{6.42}$$

6.3 ABSORBING BOUNDARY CONDITIONS

Since the FD-TDM, like the BPM in Chapter 5, has finite analysis windows, an artificial boundary condition suppressing reflections at the analysis windows is required. Mur's absorbing boundary condition (ABC) [2] is often used for this purpose, though Berenger's perfectly matched layer (PML) scheme [3] has also come into use recently. The PML scheme suppresses reflections better than Mur's ABC does, but Mur's condition is easier to use. Here, we discuss Mur's first-order ABC.

As shown in Fig. 6.2, the analysis window is defined in ranges of $(0, L_x)$ in the x direction, $(0, L_y)$ in the y direction, and $(0, L_z)$ in the z direction.

1. $x = 0$: E_y and E_z. The electric fields E_y and E_z are on the boundary $x = 0$. The wave function W for the left-traveling wave incident perpendicular to the boundary is

$$W = \exp[j(\omega t + \beta_x x)] = \exp\left[j\omega\left(t + \frac{1}{v_x}x\right)\right], \quad (6.43)$$

where v_x is the velocity of the wave. Thus, the derivatives of the wave function with respect to x and t are

$$\frac{\partial W}{\partial x} = j\omega \frac{1}{v_x} \exp\left[j\omega\left(t + \frac{1}{v_x}x\right)\right] = \frac{1}{v_x} j\omega W, \quad (6.44)$$

$$\frac{\partial W}{\partial t} = j\omega W. \quad (6.45)$$

Substituting Eq. (6.45) into (6.44), we get

$$\frac{\partial W}{\partial x} = \frac{1}{v_x} \frac{\partial W}{\partial t}$$

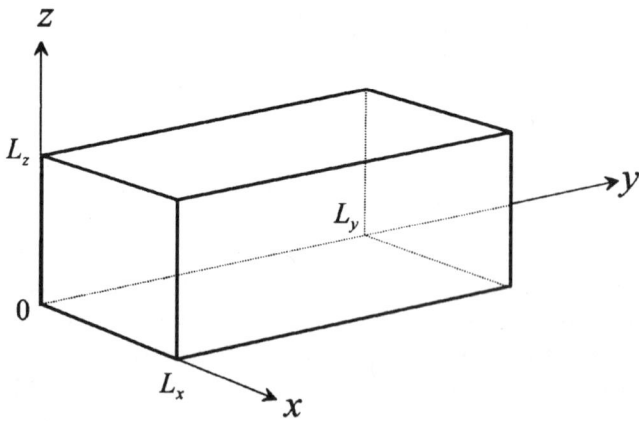

FIGURE 6.2. Analysis region.

6.3 ABSORBING BOUNDARY CONDITIONS

and therefore

$$\frac{\partial W}{\partial x} - \frac{1}{v_x}\frac{\partial W}{\partial t} = 0.$$

Thus, the wave equation for Mur's first-order ABC is

$$\left(\frac{\partial}{\partial x} - \frac{1}{v_x}\frac{\partial}{\partial t}\right)W\bigg|_{x=0} = 0. \tag{6.46}$$

Next, we discretize the wave equation (6.46) for the case that the wave function is that for the y-directed electric field E_y. Assuming that the node number of the node on the boundary is 0, we discretize the derivative of the electric field E_y with respect to the coordinate x as

$$\frac{\partial E_y}{\partial x} = \frac{1}{2}\left\{\frac{1}{\Delta x}[E_y^n(1, j+\tfrac{1}{2}, k) - E_y^n(0, j+\tfrac{1}{2}, k)]\right.$$
$$\left. + \frac{1}{\Delta x}[E_y^{n+1}(1, j+\tfrac{1}{2}, k) - E_y^{n+1}(0, j+\tfrac{1}{2}, k)]\right\}, \tag{6.47}$$

where the time average between n and $n+1$ was taken on the right-hand. On the other hand, we discretize the derivative of the electric field E_y with respect to time t as

$$\frac{\partial E_y}{\partial x} = \frac{1}{2}\left\{\frac{1}{\Delta t}[E_y^{n+1}(1, j+\tfrac{1}{2}, k) - E_y^n(1, j+\tfrac{1}{2}, k)]\right.$$
$$\left. + \frac{1}{\Delta t}[E_y^{n+1}(0, j+\tfrac{1}{2}, k) - E_y^n(0, j+\tfrac{1}{2}, k)]\right\}, \tag{6.48}$$

where the spatial average between $i = 0$ and $i = 1$ was taken. Substituting Eqs. (6.47) and (6.48) into Eq. (6.46), we can derive the following finite-difference time-domain expression for the electric field E_y:

$$E_y^{n+1}(0, j+\tfrac{1}{2}, k) = E_y^n(1, j+\tfrac{1}{2}, k)$$
$$+ \frac{v_x \Delta t - \Delta x}{v_x \Delta t + \Delta x}[E_y^{n+1}(1, j+\tfrac{1}{2}, k) - E_y^n(0, j+\tfrac{1}{2}, k)]. \tag{6.49}$$

We can similarly derive the finite-difference time-domain expression for the electric field E_z:

$$E_z^{n+1}(0,j,k+\tfrac{1}{2}) = E_z^n(1,j,k+\tfrac{1}{2})$$
$$+ \frac{v_x \Delta t - \Delta x}{v_x \Delta t + \Delta x}[E_z^{n+1}(1,j,k+\tfrac{1}{2}) - E_z^n(0,j,k+\tfrac{1}{2})]. \quad (6.50)$$

2. $x = L_x$: E_y and E_z. The electric fields E_y and E_z are on the boundary $x = L_x$. The wave function W for the right-traveling wave incident perpendicular to the boundary is

$$W = \exp[j(\omega t - \beta_x x)] = \exp\left[j\omega\left(t - \frac{1}{v_x}x\right)\right], \quad (6.51)$$

where v_x is a velocity of the wave. Thus, the derivatives of the wave function with respect to x and t are

$$\frac{\partial W}{\partial x} = -\frac{1}{v_x} j\omega W, \quad (6.52)$$

$$\frac{\partial W}{\partial t} = j\omega W. \quad (6.53)$$

Substituting Eq. (6.53) into (6.52), we get the wave equation

$$\left(\frac{\partial}{\partial x} + \frac{1}{v_x}\frac{\partial}{\partial t}\right)W\bigg|_{x=L_x} = 0. \quad (6.54)$$

Next, we discretize the wave equation (6.46) for the case that the wave function is the y-directed electric field E_y. Assuming that the node number of the node on the boundary is N_x, we discretize the derivative of the electric field E_y with respect to x as

$$\frac{\partial E_y}{\partial x} = \frac{1}{2}\left\{\frac{1}{\Delta x}[E_y^n(N_x, j+\tfrac{1}{2}, k) - E_y^n(N_x-1, j+\tfrac{1}{2}, k)]\right.$$
$$\left. + \frac{1}{\Delta x}[E_y^{n+1}(N_x, j+\tfrac{1}{2}, k) - E_y^{n+1}(N_x-1, j+\tfrac{1}{2}, k)]\right\}, \quad (6.55)$$

where the time average between n and $n+1$ was taken on the right-hand side. On the other hand, we discretize the derivative of the electric field E_y with respect to time t as

$$\frac{\partial E_y}{\partial t} = \frac{1}{2}\left\{\frac{1}{\Delta t}[E_y^{n+1}(N_x, j+\tfrac{1}{2}, k) - E_y^n(N_x, j+\tfrac{1}{2}, k)] \right.$$
$$\left. + \frac{1}{\Delta t}[E_y^{n+1}(N_x - 1, j+\tfrac{1}{2}, k) - E_y^n(N_x - 1, j+\tfrac{1}{2}, k)]\right\}, \quad (6.56)$$

where the spatial average between $i = N_x$ and $N_x - 1$ was taken. Substituting Eqs. (6.55) and (6.56) into Eq. (6.54), we can derive the finite-difference time-domain expression for the electric field E_y:

$$E_y^{n+1}(N_x, j+\tfrac{1}{2}, k) = E_y^n(N_x - 1, j+\tfrac{1}{2}, k)$$
$$+ \frac{v_x \Delta t - \Delta x}{v_x \Delta t + \Delta x}[E_y^{n+1}(N_x - 1, j+\tfrac{1}{2}, k)$$
$$- E_y^n(N_x, j+\tfrac{1}{2}, k)]. \quad (6.57)$$

We can similarly derive the finite-difference time-domain expression for the electric field E_z:

$$E_z^{n+1}(N_x, j, k+\tfrac{1}{2}) = E_z^n(N_x - 1, j, k+\tfrac{1}{2})$$
$$+ \frac{v_x \Delta t - \Delta x}{v_x \Delta t + \Delta x}[E_z^{n+1}(N_x - 1, j, k+\tfrac{1}{2})$$
$$- E_z^n(N_x, j, k+\tfrac{1}{2})]. \quad (6.58)$$

For the ABCs on $y = 0$ and $y = L_y$ and on $z = 0$ and $z = L_z$, see Problem 2.

PROBLEMS

1. Derive the restriction on the time step Δt specified in Eq. (6.41) by using the stability condition $-1 \leq A \leq 1$ in relation (6.40).

ANSWER

Using Eq. (6.33), we can rewrite the stability condition in relation (6.40) as

$$-1 \le -\frac{2(\Delta t)^2}{\varepsilon\mu_0}\frac{1}{(\Delta x)^2}\sin^2\left(\beta_x \frac{\Delta x}{2}\right) + 1 \le 1. \quad (P6.1)$$

Since the relation between the center term and right-hand term is always satisfied, we have to consider only the left-hand term and center term:

$$-1 \le -\frac{2(\Delta t)^2}{\varepsilon\mu_0}\frac{1}{(\Delta x)^2}\sin^2\left(\beta_x \frac{\Delta x}{2}\right) + 1.$$

Multiplying both sides by -1 and considering the case in which the right-hand side reaches its maximum, when $\sin^2(\cdot) = 1$, we can rewrite the above relation as

$$1 \ge \frac{2(\Delta t)^2}{\varepsilon\mu_0}\frac{1}{(\Delta x)^2} - 1.$$

Thus, we get

$$2 \ge \frac{2(\Delta t)^2}{\varepsilon\mu_0}\frac{1}{(\Delta x)^2},$$

and therefore

$$(\Delta t)^2 \le \varepsilon\mu_0 \left(\frac{1}{(\Delta x)^2}\right)^{-1}.$$

And this relation can be rewritten as the restriction on the time step Δt:

$$\Delta t \le \sqrt{\varepsilon\mu_0}\left(\frac{1}{(\Delta x)^2}\right)^{-1/2} = \frac{\sqrt{\varepsilon_r}}{c_0}\left(\frac{1}{(\Delta x)^2}\right)^{-1/2} = \frac{1}{v}\left(\frac{1}{(\Delta x)^2}\right)^{-1/2}. \quad (P6.2)$$

2. Derive the ABC fields for $y = 0$ and $y = L_y$ and $z = 0$ and $z = L_z$.

ANSWER

a. $y = 0$. The wave equation on this boundary is

$$\left(\frac{\partial}{\partial y} - \frac{1}{v_y}\frac{\partial}{\partial t}\right)W\bigg|_{y=0} = 0. \tag{P6.3}$$

The finite-difference time-domain expressions for the electric fields E_x and E_z on this boundary are

$$E_x^{n+1}(i+\tfrac{1}{2}, 0, k) = E_x^n(i+\tfrac{1}{2}, 1, k)$$
$$+ \frac{v_y \Delta t - \Delta y}{v_y \Delta t + \Delta y}[E_x^{n+1}(i+\tfrac{1}{2}, 1, k) - E_x^n(i+\tfrac{1}{2}, 0, k)], \tag{P6.4}$$

$$E_z^{n+1}(i, 0, k+\tfrac{1}{2}) = E_z^n(i, 1, k+\tfrac{1}{2})$$
$$+ \frac{v_y \Delta t - \Delta y}{v_y \Delta t + \Delta y}[E_z^{n+1}(i, 1, k+\tfrac{1}{2}) - E_x^n(i, 0, k+\tfrac{1}{2})]. \tag{P6.5}$$

b. $y = L_y$. The wave equation on this boundary is

$$\left(\frac{\partial}{\partial y} + \frac{1}{v_y}\frac{\partial}{\partial t}\right)W\bigg|_{y=L_y} = 0. \tag{P6.6}$$

The finite-difference time-domain expressions for the electric fields E_x and E_z on this boundary are

$$E_x^{n+1}(i+\tfrac{1}{2}, N_y, k) = E_x^n(i+\tfrac{1}{2}, N_y - 1, k)$$
$$+ \frac{v_y \Delta t - \Delta y}{v_y \Delta t + \Delta y}[E_x^{n+1}(i+\tfrac{1}{2}, N_y - 1, k) - E_x^n(i+\tfrac{1}{2}, N_y, k)], \tag{P6.7}$$

$$E_z^{n+1}(i, N_y, k+\tfrac{1}{2}) = E_z^n(i, N_y - 1, k+\tfrac{1}{2})$$
$$+ \frac{v_y \Delta t - \Delta y}{v_y \Delta t + \Delta y}[E_z^{n+1}(i, N_y - 1, k+\tfrac{1}{2}) - E_z^n(i, N_y, k+\tfrac{1}{2})]. \tag{P6.8}$$

c. $z = 0$. The wave equation on this boundary is

$$\left(\frac{\partial}{\partial z} - \frac{1}{v_z}\frac{\partial}{\partial t}\right)W\bigg|_{z=0} = 0.$$

The finite-difference time-domain expressions for the electric fields E_x and E_y on this boundary are

$$E_x^{n+1}(i+\tfrac{1}{2},j,0) = E_x^n(i+\tfrac{1}{2},j,1)$$
$$+ \frac{v_z \Delta t - \Delta z}{v_z \Delta t + \Delta z}[E_x^{n+1}(i+\tfrac{1}{2},j,1) - E_x^n(i+\tfrac{1}{2},j,0)],$$

(P6.9)

$$E_y^{n+1}(i,j+\tfrac{1}{2},0) = E_y^n(i,j+\tfrac{1}{2},1)$$
$$+ \frac{v_z \Delta t - \Delta y}{v_z \Delta t + \Delta y}[E_y^{n+1}(i,j+\tfrac{1}{2},1) - E_y^n(i,j+\tfrac{1}{2},0)].$$

(P6.10)

d. $z = L_z$. The wave equation on this boundary is

$$\left(\frac{\partial}{\partial z} + \frac{1}{v_z}\frac{\partial}{\partial t}\right)W\bigg|_{z=L_z} = 0. \quad (P6.11)$$

The finite-difference time-domain expressions for the electric fields E_x and E_y on this boundary are

$$E_x^{n+1}(i+\tfrac{1}{2},j,N_z) = E_x^n(i+\tfrac{1}{2},j,N_z-1)$$
$$+ \frac{v_z \Delta t - \Delta z}{v_z \Delta t + \Delta z}[E_x^{n+1}(i+\tfrac{1}{2},j,N_z-1) - E_x^n(i+\tfrac{1}{2},j,N_z)],$$

(P6.12)

$$E_y^{n+1}(i,j+\tfrac{1}{2},N_z) = E_y^n(i,j+\tfrac{1}{2},N_z-1)$$
$$+ \frac{v_z \Delta t - \Delta y}{v_z \Delta t + \Delta y}[E_y^{n+1}(i,j+\tfrac{1}{2},N_z-1) - E_y^n(i,j+\tfrac{1}{2},N_z)].$$

(P6.13)

REFERENCES

[1] K. S. Yee, "Numerical solution of initial boundary value problems involving Maxwell's equations in isotropic media," *IEEE Trans. Antennas Propagat.*, vol. AP-14, pp. 302–307, 1966.

[2] G. Mur, "Absorbing boundary conditions for the finite-difference time-domain approximation of the time domain electromagnetic field equations," *IEEE Trans. Electromagn. Compat.*, vol. EMC-23, pp. 377–382, 1981.

[3] J.-P. Berenger, "A perfectly matched layer for the absorption of electromagnetic waves," *J. Computat. Phys.*, vol. 114, pp. 185–220, 1994.

CHAPTER 7

SCHRÖDINGER EQUATION

In this chapter, we will investigate ways to solve the Schrödinger equation, which must be solved when quantum wells for semiconductor optical waveguide devices are designed [1, 2]. The time-dependent Schrödinger equation resembles the Fresnel wave equation, and the time-independent Schrödinger equation resembles the wave equation for the cross-sectional analysis. Thus, the analysis techniques given earlier for optical waveguides can be applied to the analysis of these Schrödinger equations. Here, the analysis of the time-dependent Schrödinger equation will be based on the 2D FD-BPM and the analysis of the time-independent Schrödinger equation will be based on the 1D FDM and the 1D FEM.

7.1 TIME-DEPENDENT STATE

Let us solve the time-dependent Schrödinger equation by using the 2D FD-BPM. The only major difference between the time-dependent Schrödinger equation and the BPM wave equation based on the Fresnel approximation is that the derivative in the Schrödinger equation is with respect to time, whereas the derivative in the BPM wave equation is with respect to position.

Under the effective mass approximation, the time-dependent Schrödinger equation is expressed as

$$j\hbar \frac{\partial \psi(x,t)}{\partial t} = -\frac{\hbar^2}{2} \frac{\partial}{\partial x}\left(\frac{1}{m(x)} \frac{\partial \psi(x,t)}{\partial x}\right) + U(x)\psi(x,t), \quad (7.1)$$

where $U(x)$, $m(x)$, and $\hbar$ are respectively an arbitrary potential shown in Fig. 7.1, an effective mass, and the Plank constant [3].

First, the time t and the space x are discretized. The nonequidistant discretization shown in Fig. 5.5 is assumed, and the wave function $\psi(x,t)$, the potential $U(x)$, and the effective mass $m(x)$ are expressed as

$$\psi(x,t) = \psi(x_p, t_n) = \psi_p^n, \quad (7.2)$$

$$U(x) = U(x_p) = U(p), \quad (7.3)$$

$$m(x) = m(x_p) = m(p). \quad (7.4)$$

The 2D Fresnel wave equation for the TM mode of an optical waveguide that was derived in Chapter 5 [Eq. (5.118)] is

$$2j\beta \frac{\partial \phi(x,z)}{\partial z} = \varepsilon_r(x) \frac{\partial}{\partial x}\left(\frac{1}{\varepsilon_r(x)} \frac{\partial \phi(x,z)}{\partial x}\right) + k_0^2[\varepsilon_r(x) - n_{\text{eff}}^2]\phi(x,z). \quad (7.5)$$

Comparing this with the time-dependent Schrödinger equation (7.1), we find the following correspondences:

$$z \leftrightarrow t, \quad (7.6)$$

$$\varepsilon_r(x) \frac{\partial}{\partial x}\left(\frac{1}{\varepsilon_r(x)} \frac{\partial}{\partial x}\right) \leftrightarrow -\frac{\hbar^2}{2} \frac{\partial}{\partial x}\left(\frac{1}{m(x)} \frac{\partial}{\partial x}\right), \quad (7.7)$$

$$2j\beta \leftrightarrow j\hbar, \quad (7.8)$$

$$k_0^2[\varepsilon_r(x) - n_{\text{eff}}^2] \leftrightarrow U(x). \quad (7.9)$$

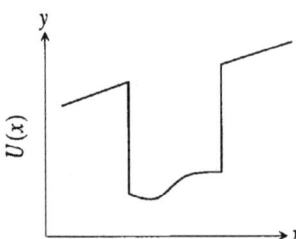

FIGURE 7.1. Potential distribution for a semiconductor quantum well.

Thus, from Eq. (5.126) and Eqs. (5.142)–(5.144), we can get the finite-difference expression for the right-hand side of correspondence (7.7):

$$-\frac{\hbar^2}{2}\frac{\partial}{\partial x}\left(\frac{1}{m(x)}\frac{\partial \phi}{\partial x}\right) = \alpha_w \psi_{p-1} + \alpha_x \psi_p + \alpha_e \psi_{p+1}, \qquad (7.10)$$

where

$$\alpha_w = -\frac{\hbar^2}{2}\frac{2}{w(e+w)}\frac{2}{m(p)+m(p-1)}, \qquad (7.11)$$

$$\alpha_e = -\frac{\hbar^2}{2}\frac{2}{e(e+w)}\frac{2}{m(p)+m(p+1)}, \qquad (7.12)$$

$$\alpha_x = \frac{\hbar^2}{2}\left(\frac{2}{w(e+w)}\frac{2}{m(p)+m(p-1)}\right.$$
$$\left. +\frac{2}{e(e+w)}\frac{2}{m(p)+m(p+1)}\right)$$
$$= -\alpha_e - \alpha_w. \qquad (7.13)$$

Considering the relations shown in correspondences (7.6)–(7.9) and in Eq. (5.135), we get the following finite-difference expression for the time-dependent Schrödinger equation:

$$-\alpha_w \psi_{p-1}^{n+1} + \left(-\alpha_x + \frac{2j\hbar}{\Delta t} - U(p)\right)\psi_p^{n+1} - \alpha_e \psi_{p+1}^{n+1}$$
$$= \alpha_w \psi_{p-1}^{l} + \left(\alpha_x + \frac{2j\hbar}{\Delta t} + U(p)\right)\psi_p^{l} + \alpha_e \psi_{p+1}^{l}. \qquad (7.14)$$

It is easily understood that this equation can be solved in the same way that the 2D FD-BPM is solved and that the transparent boundary condition can also be used with this equation.

7.2 FINITE-DIFFERENCE ANALYSIS OF TIME-INDEPENDENT STATE

The wave function $\psi(x, t)$ that satisfies the time-dependent Schrödinger equation (7.1) is divided into the space-dependent term and the time-dependent term as follows:

$$\psi(x, t) = \phi(x)\exp(-j\omega t) = \phi(x)\exp\left(-jE\frac{t}{\hbar}\right), \quad (7.15)$$

where E is the eigenenergy to be calculated and corresponds to the effective index in the optical waveguide problems.

Substituting Eq. (7.15) into Eq. (7.1) and dividing the resultant equation by $\exp(-jEt/\hbar)$, we get

$$-\frac{\hbar^2}{2}\frac{d}{dx}\left(\frac{1}{m(x)}\frac{d\phi(x)}{dx}\right) + [U(x) - E]\phi(x) = 0. \quad (7.16)$$

This is the time-independent Schrödinger equation. Discretizing it by using Eqs. (7.10)–(7.13), we get the following finite-difference expression:

$$\alpha_w\phi_{p-1} + \alpha_e\phi_{p+1} + \alpha_x\phi_p + [U(p) - E]\phi_p = 0$$

and therefore

$$\alpha_w\phi_{p-1} + [\alpha_x + U(p)]\phi_p + \alpha_e\phi_{p+1} - E\phi_p = 0. \quad (7.17)$$

The remaining task is to construct a matrix equation for Eq. (7.17).

7.3 FINITE-ELEMENT ANALYSIS OF TIME-INDEPENDENT STATE

7.3.1 Eigenvalue Equation

To solve the time-independent Schrödinger equation, we first derive the eigenvalue equation based on the FEM [4, 5]. The following normalizations are introduced:

$$\bar{x} = \frac{x}{W} \quad \text{(coordinate)}, \quad (7.18)$$

$$\bar{U}(\bar{x}) = \frac{U(x)}{E_1^\infty} \quad \text{(potential)}, \quad (7.19)$$

$$\bar{E} = \frac{E}{E_1^\infty} \quad \text{(eigenenergy)}, \quad (7.20)$$

$$\bar{m} = \frac{m(x)}{m_0} \quad \text{(effective mass)}, \quad (7.21)$$

7.3 FINITE-ELEMENT ANALYSIS OF TIME-INDEPENDENT STATE

where $E_1^\infty = \hbar^2\pi^2/(2m_0 W^2)$, m_0, and W are respectively the electron energy of the ground state in an infinitely deep well, the static mass of an electron, and the width of a quantum well.

When we substitute these normalized parameters into Eq. (7.16), we get the normalized Schrödinger equation

$$-\frac{d}{d\bar{x}}\left(\frac{1}{m(\bar{x})}\frac{d\phi(\bar{x})}{d\bar{x}}\right) + \pi^2[\bar{U}(\bar{x}) - \bar{E}]\phi(\bar{x}) = 0. \quad (7.22)$$

Next, we use the Galerkin method to solve this equation. For simplicity, the overbar, denoting normalized quantities in Eqs. (7.18) to (7.22), is omitted. Figure 7.2 shows nodes used in the 1D FEM. Using shape functions, we expand the wave function $\phi(x)$ for an element e:

$$\phi_e = \sum_{i=1}^{M_e} N_{ei}\phi_{ei} = [N_e]^T\{\phi_e\}. \quad (7.23)$$

Substituting Eq. (7.23) into Eq. (7.22), we get

$$-\frac{d}{dx}\left(\frac{1}{m_e}\frac{d}{dx}\right)[N_e]^T\{\phi_e\} + \pi^2[U(x) - E][N_e]^T\{\phi_e\} = 0.$$

Multiplying the left-hand side of this equation by the shape function and integrating the resultant equation in element e, we get

$$-\int_e [N_e]\frac{d}{dx}\left(\frac{1}{m_e}\frac{d}{dx}\right)[N_e]^T \, dx\{\phi_e\}$$

$$+ \pi^2 \int_e [N_e][U(x) - E][N_e]^T \, dx\{\phi_e\} = \{0\}. \quad (7.24)$$

Applying the partial integration to the first term of Eq. (7.24) and denoting the node numbers of the left- and right-hand nodes in element e as i and $i+1$, we rewrite Eq. (7.24) as

$$-\left[[N_e]\frac{1}{m_e}\frac{d[N_e]^T}{dx}\{\phi_e\}\right]_i^{i+1} + \int_e \frac{d[N_e]}{dx}\frac{1}{m_e}\frac{d[N_e]^T}{dx} \, dx\{\phi_e\}$$

$$+ \pi^2 \int_e [N_e](U(x) - E)[N_e]^T \, dx\{\phi_e\} = \{0\}.$$

```
     1    2    3        M-1    M
     •----•----•·········•----•
```

FIGURE 7.2. Nodes in the 1D finite-element method.

256 SCHRÖDINGER EQUATION

Furthermore, assuming that both the effective mass m and the potential $U(x)$ ($=U_e$) are constant in the element, we can reduce this equation to

$$-\left[[N_e]\frac{1}{m_e}\frac{d[N_e]^T}{dx}\{\phi_e\}\right]_i^{i+1} + \int_e \frac{1}{m_e}\frac{d[N_e]}{dx}\frac{d[N_e]^T}{dx}dx\{\phi_e\}$$
$$+\pi^2(U_e - E)\int_e [N_e][N_e]^T dx\{\phi_e\} = \{0\}. \quad (7.25)$$

Next, we sum Eq. (7.25) for all elements. The first term of Eq. (7.25) becomes

$$\sum_e \left[[N_e]\frac{1}{m}\frac{d[N_e]^T}{dx}\{\phi_e\}\right]_i^{i+1}$$
$$= \left(\phi_2 \frac{1}{m_2}\frac{d\phi_2}{dx} - \phi_1 \frac{1}{m_1}\frac{d\phi_1}{dx}\right) + \left(\phi_3 \frac{1}{m_3}\frac{d\phi_3}{dx} - \phi_2 \frac{1}{m_2}\frac{d\phi_2}{dx}\right) + \cdots$$
$$+ \left(\phi_{M-1}\frac{1}{m_{M-1}}\frac{d\phi_{M-1}}{dx} - \phi_{M-2}\frac{1}{m_{M-2}}\frac{d\phi_{M-2}}{dx}\right)$$
$$+ \left(\phi_M \frac{1}{m_M}\frac{d\phi_M}{dx} - \phi_{M-1}\frac{1}{m_{M-1}}\frac{d\phi_{M-1}}{dx}\right)$$
$$= -\phi_1 \frac{1}{m_1}\frac{d\phi_1}{dx} + \phi_M \frac{1}{m_M}\frac{d\phi_M}{dx},$$

where M is the total number of nodes. It should be noted that the following continuity conditions for the wave function and its derivative are assumed at adjacent elements e and $e+1$:

$$\phi|_e = \phi|_{e+1}, \quad \frac{1}{m}\frac{\partial\phi}{\partial x}\bigg|_e = \frac{1}{m}\frac{\partial\phi}{\partial x}\bigg|_{e+1}. \quad (7.26)$$

Thus, after summing Eq. (7.25) for all elements, we get

$$\left(\phi_1 \frac{1}{m_1}\frac{d\phi_1}{dx} - \phi_M \frac{1}{m_M}\frac{d\phi_M}{dx}\right) + \sum_e \frac{1}{m_e}\int_e \frac{d[N_e]}{dx}\frac{d[N_e]^T}{dx}dx\{\phi_e\}$$
$$+\pi^2 \sum_e U_e \int_e [N_e][N_e]^T dx\{\phi_e\} - \pi^2 E \sum_e \int_e [N_e][N_e]^T dx\{\phi_e\} = \{0\}.$$
$$(7.27)$$

7.3 FINITE-ELEMENT ANALYSIS OF TIME-INDEPENDENT STATE

This equation can be reduced to

$$\left(\phi_1 \frac{1}{m_1} \frac{d\phi_1}{dx} - \phi_M \frac{1}{m_M} \frac{d\phi_M}{dx}\right) + [P]\{\phi\} + \pi^2[Q]\{\phi\} - \pi^2 E[R]\{\phi\} = \{0\}, \tag{7.28}$$

where

$$[P] = \sum_e \frac{1}{m_e} \int_e \frac{d[N_e]}{dx} \frac{d[N_e]^T}{dx} dx, \tag{7.29}$$

$$[Q] = \sum_e U_e \int_e [N_e][N_e]^T dx, \tag{7.30}$$

$$[R] = \sum_e \int_e [N_e][N_e]^T dx, \tag{7.31}$$

$$\{\phi\} = \sum_e \{\phi_e\}. \tag{7.32}$$

Assuming the Dirichlet condition or the Neumann condition—that is, assuming

$$\phi = 0 \tag{7.33}$$

or

$$\frac{d\phi}{dx} = 0 \tag{7.34}$$

at the leftmost node 1 and the rightmost node M—we can obtain from Eq. (7.28) the simple eigenvalue equation

$$([P] + \pi^2[Q])\{\phi\} - E(\pi^2[R])\{\phi\} = \{0\}. \tag{7.35}$$

To solve this equation, we have to transform Eqs. (7.28) and (7.35) into eigenvalue matrix equations. To this end, in the following, the first- and second-order shape functions will be obtained and the explicit expressions for the matrixes will be shown.

7.3.2 Matrix Elements

A. First-Order Line Element The matrixes for the eigenvalue equation will be calculated by using the first-order line element. Figure 7.3 shows the first-order line element. The node numbers i and j and the coordinates x_i and x_j are assumed to correspond to the local coordinates 1 and 2. An arbitrary coordinate x in element e is defined using the parameter ξ, which takes a value between 0 and 1:

$$x = (1 - \xi)x_i + \xi x_j = x_i + (x_j - x_i)\xi = x_i + L_e\xi, \qquad (7.36)$$

where L_e is the length of the element $(x_j - x_i)$. Since the first-order line element has two nodes, the wave function $\phi_e(x)$ in element e is expanded as

$$\phi_e(x) = \sum_{i=1}^{2} N_i(\xi)\phi_i = [N]^T\{\phi_e\} \qquad (7.37)$$

by using the shape function $[N]^T$. The shape functions N_1 and N_2 for the line elements are expressed by the linear functions

$$N_1(\xi) = a_1\xi + b_1, \qquad N_2(\xi) = a_2\xi + b_2, \qquad (7.38)$$

and the conditions that must be met by the shape functions are

$$\xi = 0: \qquad N_1(0) = 1, \qquad N_2(0) = 0, \qquad (7.39)$$
$$\xi = 1: \qquad N_1(1) = 0, \qquad N_2(1) = 1. \qquad (7.40)$$

Thus, the following shape functions can be obtained for the first-order line element:

$$N_1(\xi) = 1 - \xi, \qquad (7.41)$$
$$N_2(\xi) = \xi. \qquad (7.42)$$

Next, we calculate the matrix elements shown in Eqs. (7.29)–(7.32).

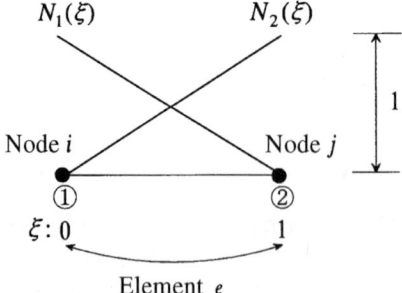

FIGURE 7.3. First-order line element.

7.3 FINITE-ELEMENT ANALYSIS OF TIME-INDEPENDENT STATE

1. $\int_e (d[N_e]/dx)(d[N_e]^T/dx)\, dx$ From Eq. (7.36), we get

$$\frac{d\xi}{dx} = \frac{1}{L_e}$$

and therefore,

$$dx = L_e\, d\xi. \tag{7.43}$$

Since the relations

$$\frac{dN_1}{d\xi} = -1, \tag{7.44}$$

$$\frac{dN_2}{d\xi} = 1, \tag{7.45}$$

hold, we get

$$\frac{d[N_e]}{dx} = \frac{d\xi}{dx}\frac{d[N_e]}{d\xi} = \frac{1}{L_e}\begin{bmatrix} -1 \\ 1 \end{bmatrix}. \tag{7.46}$$

Thus, we get

$$\int_e \frac{d[N_e]}{dx}\frac{d[N_e]^T}{dx}\, dx = \frac{1}{L_e^2}\int_0^1 \begin{bmatrix} -1 \\ 1 \end{bmatrix}[-1\ \ 1]L_e\, d\xi$$

$$= \frac{1}{L_e}\int_0^1 \begin{bmatrix} 1 & -1 \\ -1 & 1 \end{bmatrix} d\xi$$

$$= \frac{1}{L_e}\begin{bmatrix} 1 & -1 \\ -1 & 1 \end{bmatrix}. \tag{7.47}$$

2. $\int_e [N_e][N_e]^T\, dx$ Through a similar procedure, we get

$$\int_e [N_e][N_e]^T\, dx = \int_0^1 \begin{bmatrix} (1-\xi) \\ \xi \end{bmatrix}[(1-\xi)\ \ \xi]L_e\, d\xi$$

$$= L_e \int_0^1 \begin{bmatrix} (1-\xi)^2 & \xi(1-\xi) \\ \xi(1-\xi) & \xi^2 \end{bmatrix} d\xi = \frac{L_e}{6}\begin{bmatrix} 2 & 1 \\ 1 & 2 \end{bmatrix}. \tag{7.48}$$

Since Eqs. (7.47) and (7.48) can be used to construct the matrixes $[P]$, $[Q]$, and $[R]$, the eigenenergy can be calculated from the eigenvalue matrix equation (7.35).

B. Second-Order Line Element Next, we discuss the second-order line element, which is more accurate than the first-order line element. Figure 7.4 shows the second-order line element. The node numbers i, j, and k and the coordinates x_i, x_j, and x_k are assumed to correspond to the local coordinates 1, 2, and 3. An arbitrary coordinate x in element e is defined using the parameter ξ, which takes a value between -1 and 1:

$$x = x_j + \tfrac{1}{2}(x_k - x_i)\xi. \tag{7.49}$$

Since the second-order line element has three nodes, the wave function $\phi_e(x)$ in element e is expanded as

$$\phi_e(x) = \sum_{i=1}^{3} N_i(\xi)\phi_i = [N_e]^T \{\phi_e\} \tag{7.50}$$

by using the shape function $[N]^T$. The shape functions N_1, N_2, and N_3 for the line elements are expressed by the quadratic polynomials

$$N_1(\xi) = a_1\xi^2 + b_1\xi + c_1,$$
$$N_2(\xi) = a_2\xi^2 + b_2\xi + c_2, \tag{7.51}$$
$$N_3(\xi) = a_3\xi^2 + b_3\xi + c_3,$$

and the conditions that must be met by the shape functions are

$\xi = 0$: $N_1(0) = 0$, $N_2(0) = 1$, $N_3 = 0$,   (7.52)

$\xi = 1$: $N_1(1) = 0$, $N_2(1) = 0$, $N_3(1) = 1$,   (7.53)

$\xi = -1$: $N_1(-1) = 1$, $N_2(-1) = 0$, $N_3(-1) = 0$.   (7.54)

Thus, the following shape functions can be obtained for the second-order line element:

$$N_1(\xi) = -\tfrac{1}{2}\xi(1 - \xi), \tag{7.55}$$

$$N_2(\xi) = (1 + \xi)(1 - \xi), \tag{7.56}$$

$$N_3(\xi) = \tfrac{1}{2}\xi(1 + \xi). \tag{7.57}$$

Next, we calculate the matrix elements shown in Eqs. (7.29)–(7.32).

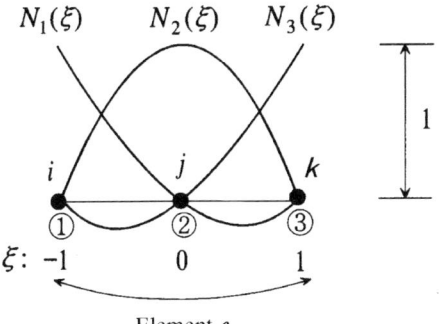

FIGURE 7.4. Second-order line element.

1. $\int_e (d[N_e]/dx)(d[N_e]^T/dx)\, dx$ From Eq. (7.49), we get

$$\frac{d\xi}{dx} = \frac{2}{L_e}$$

and therefore,

$$dx = \tfrac{1}{2} L_e\, d\xi. \tag{7.58}$$

Since the relations

$$\frac{dN_1}{d\xi} = \xi - \tfrac{1}{2}, \tag{7.59}$$

$$\frac{dN_2}{d\xi} = -2\xi, \tag{7.60}$$

$$\frac{dN_3}{d\xi} = \xi + \tfrac{1}{2} \tag{7.61}$$

hold, we get

$$\frac{d[N_e]}{dx} = \frac{d\xi}{dx}\frac{d[N_e]}{d\xi} = \frac{2}{x_k - x_i}\begin{bmatrix} \xi - \tfrac{1}{2} \\ -2\xi \\ \xi + \tfrac{1}{2} \end{bmatrix} = \frac{1}{x_k - x_i}\begin{bmatrix} 2\xi - 1 \\ -4\xi \\ 2\xi + 1 \end{bmatrix}. \tag{7.62}$$

262 SCHRÖDINGER EQUATION

Thus, we get

$$\int_e \frac{d[N_e]}{dx} \frac{d[N_e]^T}{dx} dx$$

$$= \frac{1}{(x_k - x_i)^2} \int_{-1}^{1} \begin{bmatrix} 2\xi - 1 \\ -4\xi \\ 2\xi + 1 \end{bmatrix} [2\xi - 1 \quad -4\xi \quad 2\xi + 1] \frac{(x_k - x_i)}{2} d\xi$$

$$= \frac{1}{2(x_k - x_i)}$$

$$\times \int_{-1}^{1} \begin{bmatrix} (2\xi - 1)^2 & (2\xi - 1)(-4\xi) & (2\xi - 1)(2\xi + 1) \\ -4\xi(2\xi - 1) & (-4\xi)^2 & (-4\xi)(2\xi + 1) \\ (2\xi - 1)(2\xi + 1) & (2\xi + 1)(-4\xi) & (2\xi + 1)^2 \end{bmatrix} d\xi$$

$$= \frac{1}{2(x_k - x_i)} \frac{1}{3} \begin{bmatrix} 14 & -16 & 2 \\ -16 & 32 & -16 \\ 2 & -16 & 14 \end{bmatrix}.$$

This equation is summarized as

$$\int_e \frac{d[N_e]}{dx} \frac{d[N_e]^T}{dx} dx = \frac{1}{6L_e} \begin{bmatrix} 14 & -16 & 2 \\ -16 & 32 & -16 \\ 2 & -16 & 14 \end{bmatrix}. \qquad (7.63)$$

2. $\int_e [N_e][N_e]^T dx$ Through a similar procedure, we get

$$\int_e [N_e][N_e]^T dx$$

$$= \int_{-1}^{1} \begin{bmatrix} -\frac{1}{2}\xi(1 - \xi) \\ (1 + \xi)(1 - \xi) \\ \frac{1}{2}\xi(1 + \xi) \end{bmatrix} [-\frac{1}{2}\xi(1 - \xi) \quad (1 + \xi)(1 - \xi) \quad \frac{1}{2}\xi(1 + \xi)]$$

$$\times [\tfrac{1}{2}(x_k - x_i)] d\xi$$

$$= \frac{L_e}{2} \int_{-1}^{1} \begin{bmatrix} \frac{1}{4}\xi^2(1 - \xi)^2 & -\frac{1}{2}\xi(1 + \xi)(1 - \xi)^2 & -\frac{1}{4}(1 + \xi)(1 - \xi) \\ -\frac{1}{2}\xi(1 - \xi)^2(1 + \xi) & (1 + \xi)^2(1 - \xi)^2 & \frac{1}{2}\xi(1 + \xi)^2(1 - \xi) \\ -\frac{1}{4}\xi^2(1 + \xi)(1 - \xi) & \frac{1}{2}\xi(1 + \xi)^2(1 - \xi) & \frac{1}{4}\xi^2(1 + \xi)^2 \end{bmatrix} d\xi$$

$$= \frac{L_e}{2} \frac{1}{15} \begin{bmatrix} 4 & 2 & -1 \\ 2 & 16 & 2 \\ -1 & 2 & 4 \end{bmatrix} = \frac{L_e}{30} \begin{bmatrix} 4 & 2 & -1 \\ 2 & 16 & 2 \\ -1 & 2 & 4 \end{bmatrix}. \qquad (7.64)$$

Since Eqs. (7.63) and (7.64) can be used to construct the matrixes $[P]$, $[Q]$, and $[R]$, the eigenenergy can be calculated from the eigenvalue matrix equation (7.35).

REFERENCES

[1] K. Kawano, S. Sekine, H. Takeuchi, M. Wada, M. Kohtoku, N. Yoshimoto, T. Ito, M. Yanagibashi, S. Kondo, and Y. Noguchi, "4 × 4 InGaAlAs/InAlAs MQW directional coupler waveguide switch modules integrated with spot-size converters and their 10 Gbit/s operation," *Electron. Lett.*, vol. 31, pp. 96–97, 1995.

[2] K. Kawano, K. Wakita, O. Mitomi, I. Kotaka, and M. Naganuma, "Design of InGaAs/InAlAs multiple-quantum well (MQW) optical modulators," *IEEE J. Quantum Electron.*, vol. QE-28, pp. 228–230, 1992.

[3] L. I. Schiff, *Quantum Mechanics*, McGraw-Hill, New York, 1968.

[4] K. Nakamura, A. Shimizu, M. Koshiba, and K. Hayata, "Finite-element analysis of quantum wells of arbitray semiconductors with arbitrary potential potential profiles," *IEEE J. Quantum Electron.*, vol. 25, pp. 889–895, 1989.

[5] M. Koshiba, H. Saitoh, M. Eguchi, and K. Hirayama, 'Simple scalar finite-element approach to optical waveguides," *IEE Proc. J.*, vol. 139, pp. 166–171, 1992.

APPENDIX A

VECTORIAL FORMULAS

In the following, **i**, **j** and **k** are respectively unit vectors in the x, y, and z directions and ϕ and **A** are respectively a scalar and a vector:

$$\mathbf{A} = A_x\mathbf{i} + A_y\mathbf{j} + A_z\mathbf{k}, \tag{A.1}$$

$$\mathbf{\nabla} = \frac{\partial}{\partial x}\mathbf{i} + \frac{\partial}{\partial y}\mathbf{j} + \frac{\partial}{\partial z}\mathbf{k}, \tag{A.2}$$

$$\mathbf{\nabla} \cdot \mathbf{\nabla} \times \mathbf{A} = 0, \tag{A.3}$$

$$\mathbf{\nabla} \times (\mathbf{\nabla} \times \mathbf{A}) = \mathbf{\nabla}(\mathbf{\nabla} \cdot \mathbf{A}) - \nabla^2 \mathbf{A}, \tag{A.4}$$

$$\nabla^2 = \frac{\partial^2}{\partial x^2} + \frac{\partial^2}{\partial y^2} + \frac{\partial^2}{\partial z^2}, \tag{A.5}$$

$$\nabla_\perp^2 = \frac{\partial^2}{\partial x^2} + \frac{\partial^2}{\partial y^2}, \tag{A.6}$$

$$\mathbf{\nabla}(\phi\mathbf{A}) = \mathbf{\nabla}\phi \cdot \mathbf{A} + \phi\mathbf{\nabla} \cdot \mathbf{A}, \tag{A.7}$$

$$\mathbf{\nabla} \times \mathbf{A} = \begin{vmatrix} \mathbf{i} & \mathbf{j} & \mathbf{k} \\ \frac{\partial}{\partial x} & \frac{\partial}{\partial y} & \frac{\partial}{\partial z} \\ A_x & A_y & A_z \end{vmatrix}$$

$$= \left(\frac{\partial A_z}{\partial y} - \frac{\partial A_y}{\partial z}\right)\mathbf{i} + \left(\frac{\partial A_x}{\partial z} - \frac{\partial A_z}{\partial x}\right)\mathbf{j} + \left(\frac{\partial A_y}{\partial x} - \frac{\partial A_x}{\partial y}\right)\mathbf{k}. \tag{A.8}$$

If **r**, **θ**, and **z** are respectively unit vectors in the radial, azimuthal, and longitudinal directions, the rotation formula for a vector $\mathbf{A} = A_r \mathbf{r} + A_\theta \boldsymbol{\theta} + A_z \mathbf{z}$ is expressed as

$$\nabla \times \mathbf{A} = \left(\frac{1}{r} \frac{\partial A_z}{\partial \theta} - \frac{\partial A_\theta}{\partial z} \right) \mathbf{r} + \left(\frac{\partial A_r}{\partial z} - \frac{\partial A_z}{\partial r} \right) \boldsymbol{\theta} + \left(\frac{1}{r} \frac{\partial}{\partial r}(r A_\theta) - \frac{1}{r} \frac{\partial A_r}{\partial \theta} \right) \mathbf{z}. \tag{A.9}$$

A Laplacian ∇^2 for a cylindrical coordinate is given as

$$\begin{aligned}
\nabla^2 &= \nabla_\perp^2 + \frac{\partial^2}{\partial z^2} \\
&= \frac{1}{r} \frac{\partial}{\partial r} \left(r \frac{\partial}{\partial r} \right) + \frac{1}{r^2} \frac{\partial^2}{\partial \theta^2} + \frac{\partial^2}{\partial z^2} \\
&= \frac{\partial^2}{\partial r^2} + \frac{1}{r} \frac{\partial}{\partial r} + \frac{1}{r^2} \frac{\partial^2}{\partial \theta^2} + \frac{\partial^2}{\partial z^2}.
\end{aligned} \tag{A.10}$$

APPENDIX B

INTEGRATION FORMULA FOR AREA COORDINATES

The integration formula shown in Eq. (3.184) is derived here by calculating the following integration for a triangular element e shown in Fig. 3.4:

$$I_e(i,j,k) = \iint_e L_1^i L_2^j L_3^k \, dx \, dy, \tag{B.1}$$

where i, j, and k are integers and the spatial coordinates of nodes 1, 2, and 3 are respectively (x_1, y_1), (x_2, y_2), and (x_3, y_3).

As shown in Eq. (3.72), we have the following relation between the spatial coordinates and the area coordinates:

$$\begin{pmatrix} x \\ y \\ 1 \end{pmatrix} = \begin{pmatrix} x_1 & x_2 & x_3 \\ y_1 & y_2 & y_3 \\ 1 & 1 & 1 \end{pmatrix} \begin{pmatrix} L_1 \\ L_2 \\ L_3 \end{pmatrix}. \tag{B.2}$$

As shown by the bottom row of Eq. (B.2),

$$L_1 + L_2 + L_3 = 1. \tag{B.3}$$

INTEGRATION FORMULA FOR AREA COORDINATES

According to Eq. (B.2), x and y can be expressed using L_1 and L_2 as follows:

$$\begin{aligned} x &= x_1 L_1 + x_2 L_2 + x_3 L_3 \\ &= x_1 L_1 + x_2 L_2 + x_3(1 - L_1 - L_2) \\ &= (x_1 - x_3) L_1 + (x_2 - x_3) L_2, \end{aligned} \quad \text{(B.4)}$$

$$\begin{aligned} y &= y_1 L_1 + y_2 L_2 + y_3 L_3 \\ &= y_1 L_1 + y_2 L_2 + y_3(1 - L_1 - L_2) \\ &= (y_1 - y_3) L_1 + (y_2 - y_3) L_2. \end{aligned} \quad \text{(B.5)}$$

Thus, we get

$$\frac{\partial(x, y)}{\partial(L_1, L_2)} = \begin{vmatrix} \frac{\partial x}{\partial L_1} & \frac{\partial x}{\partial L_2} \\ \frac{\partial y}{\partial L_1} & \frac{\partial y}{\partial L_2} \end{vmatrix} = \begin{vmatrix} x_1 - x_3 & x_2 - x_3 \\ y_1 - y_3 & y_2 - y_3 \end{vmatrix} = 2S_e, \quad \text{(B.6)}$$

where S_e is the area of element e. Using Eq. (B.6) to transform x and y to L_1 and L_2, we can rewrite Eq. (B.1) as follows:

$$\begin{aligned} I_e(i, j, k) &= \iint_e L_1^i L_2^j L_3^k \, dx \, dy \\ &= \iint_e L_1^i L_2^j L_3^k \left| \frac{\partial(x, y)}{\partial(L_1, L_2)} \right| dx \, dy \\ &= 2S_e \iint_e L_1^i L_2^j L_3^k \, dL_1 \, dL_2 \\ &= 2S_e \int_0^1 dL_1 \int_0^{1-L_1} L_1^i L_2^j (1 - L_1 - L_2)^k \, dL_2 \\ &= 2S_e \int_0^1 L_1^i \, dL_1 \int_0^{1-L_1} L_2^j (1 - L_1 - L_2)^k \, dL_2. \end{aligned} \quad \text{(B.7)}$$

The second integral of Eq. (B.7) is calculated as

$$I_0(j, k) = \int_0^{1-L_1} L_2^j (1 - L_1 - L_2)^k \, dL_2$$

$$= \left[\frac{1}{j+1} L_2^{j+1} (1 - L_1 - L_2)^k \right]_0^{1-L_1}$$

$$+ \frac{k}{j+1} \int_0^{1-L_1} L_2^{j+1} (1 - L_1 - L_2)^{k-1} \, dL_2$$

$$= 0 + \frac{k}{j+1} \int_0^{1-L_1} L_2^{j+1} (1 - L_1 - L_2)^{k-1} \, dL_2$$

$$= \frac{k}{j+1} I_0(j+1, k-1)$$

$$= \left[\frac{1}{j+2} L_2^{j+2} (1 - L_1 - L_2)^{k-1} \right]_0^{1-L_1}$$

$$+ \frac{k(k-1)}{(j+1)(j+2)} \int_0^{1-L_1} L_2^{j+2} (1 - L_1 - L_2)^{k-2} \, dL_2$$

$$= 0 + \frac{k(k-1)}{(j+1)(j+2)} I_0(j+2, k-2)$$

$$= \frac{k(k-1) \cdots 1}{(j+1)(j+2) \cdots (j+k)} I_0(j+k, 0)$$

$$= \frac{k! j!}{(j+k)!} I_0(j+k, 0), \tag{B.8}$$

where

$$I_0(j+k, 0) = \int_0^{1-L_1} L_2^{j+k} \, dL_2 = \left[\frac{1}{j+k+1} L_2^{j+k+1} \right]_0^{1-L_1}$$

$$= \frac{1}{j+k+1} (1 - L_1)^{j+k+1}. \tag{B.9}$$

Substituting Eq. (B.9) into Eq. (B.8), we get

$$I_0(j, k) = \frac{j! k!}{(j+k+1)!} (1 - L_1)^{j+k+1}. \tag{B.10}$$

And substituting Eq. (B.10) into Eq. (B.7), we get

$$I_e(i,j,k) = 2S_e \frac{j!k!}{(j+k+1)!} \int_0^1 L_1^i (1-L_1)^{j+k+1} \, dL_1. \tag{B.11}$$

The remaining calculation is

$$I_1(i, j+k+1) = \int_0^1 L_1^i (1-L_1)^{j+k+1} \, dL_1$$

$$= \left[\frac{1}{i+1} L_1^{i+1} (1-L_1)^{j+k+1} \right]_0^1$$

$$+ \frac{j+k+1}{i+1} \int_0^1 L_1^{i+1} (1-L_1)^{j+k} \, dL_1$$

$$= 0 + \frac{j+k+1}{i+1} I_1(i+1, j+k)$$

$$= \frac{(j+k+1)(j+k) \cdots 2 \cdot 1}{(i+1)(i+2) \cdots (i+j+k+1)} I_1(i+j+k+1, 0)$$

$$= \frac{(j+k+1)! i!}{(i+j+k+1)!} I_1(i+j+k+1, 0). \tag{B.12}$$

Substituting $I_1(i+j+k+1, 0)$, where $I_1(i+j+k+1, 0)$ can be rewritten as

$$I_1(i+j+k+1, 0) = \int_0^1 L_1^{i+j+k+1} \, dL_1 = \left[\frac{1}{i+j+k+2} L_1^{i+j+k+2} \right]_0^1$$

$$= \frac{1}{i+j+k+2}, \tag{B.13}$$

into Eq. (B.12), we get

$$I_1(i, j+k+1) = \frac{(j+k+1)! i!}{(i+j+k+2)!}. \tag{B.14}$$

Then substituting this equation into Eq. (B.11), we finally get the formula

$$I_e(i, j, k) = 2S_e \frac{i!\, j!\, k!}{(i + j + k + 2)!}. \tag{B.15}$$

INDEX

Area coordinate, 74
Alternate-direction implicit (ADI) method, 216
Angular frequency, 3

Bandwidth, 109
Basis function, 62
Beam propagation method (BPM), 165
    ADI-BPM, 216
    first Fourier transform beam propagation method (FFT-BPM), 165
    finite-difference beam propagation method (FD-BPM), 180
Bessel function, 40
    Bessel function of first kind, 40
    Bessel function of second kind, 40
    modified Bessel function of first kind, 40
    modified Bessel function of second kind, 40
Boundary condition, 9, 27, 110, 153, 197
    absorbing boundary condition (ABC), 241
    analytical boundary condition, 154
    transparent boundary condition (TBC), 197

Characteristic equation, 17, 20, 41, 53
Charge density, 1
Cladding, 13, 37, 125
Core, 13, 37, 125
Cramer's formula, 76
Crank-Nicolson scheme, 195
Current density, 1
Cylindrical coordinate system, 38

Derivative, 119
    first derivative, 119
    second derivative, 119
Difference center, 131
    hypothetical difference center, 131
Discretization, 130
    equidistant discretization, 130
    nonequidistant discretization, 130
Dirichlet condition, 66, 111, 153, 257
Dominant mode, 111

Effective index, 6
    effective index method, 20
Eigenvalue, 88, 151
    eigenvalue matrix equation, 68, 72, 151, 257
Eigenvector, 88, 151
Electric field, 1

Electric flux density, 1
Element, 64
　triangular element, 64, 72
　first-order triangular element, 64, 73, 91, 106
　second-order triangular element, 64, 79, 95, 108
E^x_{pq} mode, 24, 113
E^y_{pq} mode, 31, 113
Even mode, 111
Expansion coefficient, 63
Explicit scheme, 239

Finite-element method (FEM), 59
　scalar finite-element method (SC-FEM), 59
Finite-difference method (FDM), 116
　scalar finite-difference method (SC-FDM), 150
　semivectorial finite-difference method (SV-FDM), 117
Finite-difference time-domain method (FD-TDM), 233
Fourier transform, 170
　discrete Fourier transform, 170
　inverse discrete Fourier transform, 170

Fresnel approximation, 167–168, 187
Functional, 62
Fully vectorial analysis, 222

Galerkin method, 68

Helmholtz equaiotn, 5–7
Hybrid-mode analysis, 47

Implicit scheme, 186
Interpolation function, 64

Joule heating, 8

Laplacian, 4
Line element, 257
　first-order line element, 257
　second-order line element, 260
Local coordinate, 107, 110
LP mode, 38

Magnetic field, 1
Magnetic flux density, 1
Marcatili's method, 23
Maxwell's equations, 1
Matrix element, 89
Mirror-symmetrical plane, 111
Multistep method, 213

Neumann condition, 66, 111, 153, 257
Node, 64, 73, 79, 129, 257
Normalized frequency, 41

Odd mode, 111
Optical fiber, 36
　step-index optical fiber, 37

Padé approximant operator, 204
Para-axial approximation, 167
Permeability, 1
　relative permeability, 1
Permittivity, 1
　relative permittivity, 1
Phase-shift lens, 170, 173
Phasor expression, 3
Plane wave, 10
Plank constant, 252
Potential, 252
Power confinement factor (Γ factor), 55–56
Poynting vector, 7
Principal field component, 125, 182, 184
Propagation constant, 6

Quantum well, 252
Quasi-TE mode, 125, 128
Quasi-TM mode, 125, 147

Rayleigh-Ritz method, 62
Reference index, 166, 187
Residual, 68
　error residual, 68
　weighted residual method, 69

Schrödinger equation, 251
　normalized Schrödinger equation, 255
　time-dependent Schrödinger equation, 251

time-independent Schrödinger equation, 253–254
Shape function, 64, 78, 83
Slab optical waveguide, 13
Slowly varying envelope approximation (SVEA), 166
Stability condition, 195, 239

Taylor series expansion, 118
Tohmas method, 203
Transverse electric (TE) mode, 14, 181, 186
Transverse magnetic (TM) mode, 14, 184, 190

Variational method, 59
Variational principle, 62

Wave equation, 5, 6
  scalar wave equation, 84, 127
  semivectorial wave equation, 124
  vectorial wave equation, 4, 120
Wave number, 5
Weak form, 69
Wide-angle formulation, 167
Wide-angle analysis, 204
Wide-angle order, 205

Yee lattice, 235